Celebrating Humanity

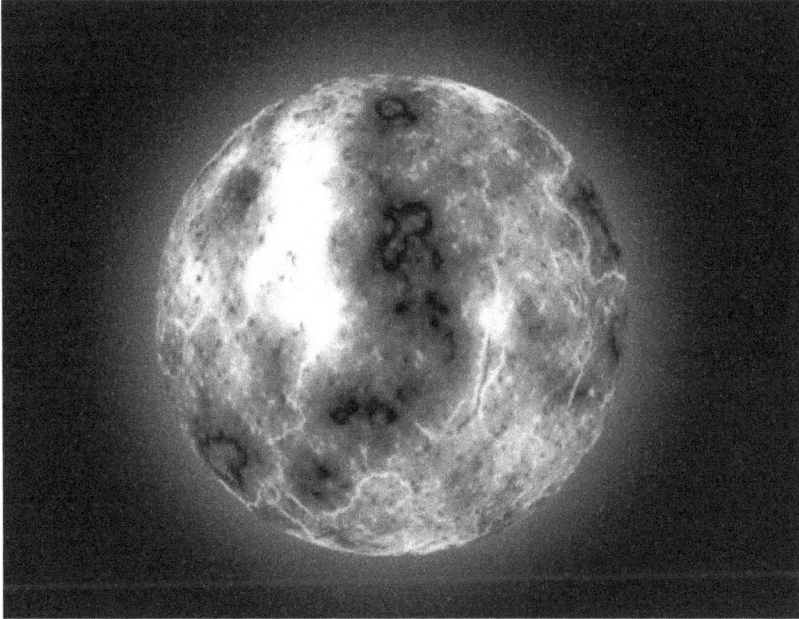

A flaming Earth approximately 2 billion years ago

A comprehensive analysis of the Human Experience, one that includes the Pain, The Glory and the Complexities faced by the members of a rather unique specie.

Dr. M. Franklin

The Center for Intellectual Development

ISBN   9781627190152
Celebrating Humanity
Milton Franklin, Author
Yolanda Franklin, Editor
Published by Vizcaya International Inc.

Table of Contents

DEDICATION

This book is dedicated to the men and women of all ages who dare to think, standing up to the onslaught of intimidation perpetually unleashed by the mindless dogma driven individuals in all countries. The simple yet powerful statement being made by this writing is that the era of intimidation is over. The gathering of thinkers that this book represents is a process that makes no apologies for the ideas it advances, and it will stop at nothing to empower those who dare to think.

Thinkers of all ages willing to practice intellectual honesty by going where their thoughts take them will find a welcoming home in our organization, The Center for Intellectual Development. It is the organized body on which we depend for carrying forth the simple idea that the loud mouths and intimidators no longer control the conversation between humans. In short, it is written for the young men and women across the globe that are curious enough to want to know about all that is worthwhile knowing.

In purchasing this book you've made a leap of faith, one that will transform you from an ordinary human being into a clear minded person with the vigor and vitality to alter history for the better

INTRODUCTION

I can't recall in this my long life a period as complex as the one in which we live today. H hardly anything makes sense and people are really at a lost for meaning in their lives. Just when I thought my crusading years were all behind me, I'm faced with a dilemma that compels me to speak out not on behalf of anyone in particular, but in asserting a reality that nearly everyone I know chooses to ignore.

At the core of my thinking and at the heart of everything I do at this stage in my life is the notion that this species; almost in its totality; will allow itself to be tripped up by mythology and childish thinking even after the extraordinary scientific revelations of the past two hundred years. The most emblematic of this childish thinking are the vast number of humans denouncing science while clutching their cellular telephone one of the greatest achievements in science of all times. There have been other significant achievements in science in our times but none hold sway over us as a specie as the cellular telephone has.

The front cover depiction of planet earth approximately 2 billion years ago is at the core of all of the manifestations made in these pages, and as painful as it is the separation is between those who can conceptualize this reality, and those who refuse to or are unable to embrace this complex reality. The embracing of this reality puts aside all notions of God or Jesus or Mohammed or Allah, or Buda, or the gazillion other deities invented by humans, and the dangerous belief systems built around them.

Four hundred years ago research conducted by the Polish scientist Nicolas Copernicus revealed to the world that the Sun does not revolve around our planet. Those findings were suppressed however, as the powerful religious institutions of his time would have executed for going against the sacred edicts of the church. Much more has been revealed since then but unfortunately these extraordinary scientific revelations have not made their way into the subconscious of the overwhelming majority of humans. That precisely is the reasoning behind this new proposal. The term Scientific Psychology is thus introduced to describe the relationship that should exist between the scientific discoveries in all of the ages, and the practice of psychology, or more specifically, psychotherapy. My own experience in more than three decades of psychotherapy says that the introduction of this factual evidentiary information has the potential of inoculating the individual against delusion and against the massive confusion that forever engulfs the masses.

Since time immemorial, scientific researchers have been bringing us information about ourselves and the world in which we live, and the cosmological scenery that is presented with this information.

This information of course, is far removed from nearly everything humans have been told by the clergy of various stripes and denominations. The self-discovery it offers holds promise for the future as the stranglehold that religion continues to have on the mind of the masses is lessened. The ignorance, the prejudice, and the racism that is supported by narrow religious teachings will also decrease as these findings make it into the subconscious.

Thanks to scientists the psychological playing field has been leveled, despite the fact that the information has not gotten to all quarters. It is therefore only a matter of time as precise information about our common stock and our common origins as a species take hold with new generations of humans on this planet.

The Center for Intellectual Development, a not for profit organization legally registered in the State of Florida and in Costa Rica, has had great success in the use of scientific psychology both in its therapeutic interventions and in its educational seminars. It is empowering because it removes the notion of victim that most survivors of abuse carry, and it prepares them to take charge of their own lives as they assert themselves in a world that is distinctively theirs.

It is not rocket science, in fact the method is alarmingly simple. The only difficulty facing this new method is the hypnotic and narcotizing effect the various religious forms has had on the overwhelming majority of the population. For those who are no longer under the spell of religion this method has far reaching effects.

My partiality towards the intellect is one I could not shake even if I wanted to, for I've read and witnessed too much unnecessary sufferings purely on account of its absence. I may succeed at persuading only a few to the cause of intellectual development, but those few are a cause for great celebration given the reality of our times. I will be bringing this organization to anyone open enough to listen, and in so doing branches will be developed in any corner of the world where people are willing to relate to each other with open-mindedness, with love, and with compassion.

As stated, much has happened since that first attempt where, my eternal optimism leads me to focus on the subject of human happiness. I was concerned about the pain and confusion in which so many people live their lives, but half way through it I realized that a lot of this suffering is self-inflicted not realizing fully that the one thing that is needed to overcome a state of poverty and misery is information. However, for reasons that I find difficult to explain, people in general prefer to remain uninformed about the most vital issues concerning their lives. They prefer to remain infantile, believing that there is a father figure somewhere beyond the skies who sooner or later will make everything right. They refuse to become involved in the hard work that is involved in acquiring information, and they are afraid that being informed will inevitably force them to act. So being uninformed makes everything easier. Whatever the reason, people in general chose to remain ignorant. It has become their comfort zone, and despite their complaining, they're happy in that zone. This constituted the ultimate education for me but I couldn't turn back, I was already committed, so I was forced to alter my course of action. This could no longer be a book about happiness but rather, one that identifies and deals with the cause of misery. True to self I had to point out ignorance and stupidity as the primary cause of most human suffering. This will no doubt irate a great many, but there is simply no other way of putting it. In an era so replete with information one has to make an effort to remain ignorant, regrettably most people are making that effort. I'm not suggesting that knowledge alone will cure all human suffering, for given the work that I do I'm aware that there are those who are genuinely ill, struggling to overcome the traumas inflicted on

4

them when they were only babies of four, and seven and 10 years old, some even older. My heart goes out to those, and it is for them that this book is written. My associates and I are determined to create a place where these suffering individuals; most of whom are extremely talented; can find solace, can find peace comfort and strength. It is not for the Smart Alex' who have all the answers; in fact, it is not even recommended that they read this book for it would do them no good. The **Center for Intellectual Development** has created a place for all those who wish to free themselves from their own suffering. That is our only goal. But one cannot be a part of this process while holding on to the very things that has made their lives miserable, something must give. In this process everything we've ever held sacred is up for scrutiny, and if it doesn't pass the test it will be abolished. Culture and belief systems are at the core of everyone's existence, yet few of us know what they are. If we had to give a simple and short definition for what we've come to regard as Culture, it would be **learned behavior**. Belief, unfortunately, is something we are forced to do before we even learn how to think. Wherever belief supersedes thinking that society is doomed. One only has to look at the socio-economic state of Black People today to realize what happens when a people place belief above thinking. They were given a set of beliefs by their slave masters, and they continue to clutch these long after the master had set them free. They fail to realize that they were only set free because the master no longer had use for them, he had developed machinery that could do the work faster and more efficiently. Their efforts had little or nothing to do with the master's decision to set them free although he has allowed them to think that it was their struggle

5

for liberation that made it happen. I suppose some dignity was preserved by allowing them to think that, but the delusion it encouraged also caused permanent damage in the psyche of these ex-slaves for it almost forced them to see the master as benevolent and full of compassion. This delusion of course, causes them to be eternally grateful to their former masters, and nothing could be more damaging.

I have great respect for those who dedicate themselves to this lonely and often unrewarding profession. I became involved in this business of writing only because no one else is willing to say the things that I find to be compelling. I've learned a great deal in the process but I would just as soon have someone else do the writing while I dedicate time to my real passions that are: building, cooking, mentoring and organizing. I organize because of the love I have for my fellow humans, and this undying love has made a scholar out of me and the driving force behind my desire to learn all that I can about self and others. It is why I have ventured into the field of history, psychology and philosophy. The ultimate objective of the organization is the emotional and economic stability of all its members. All children are considered honorary members of this organization regardless of their parent's mindset. We will forever fight for their rights to be treated with love, kindness and respect regardless of their behavior.

We've decided in this organization that there is only one enemy to be fought against, and that is *ignorance*, none other, and as far as we are concerned the only weapons needed in this battle are: ***Wisdom, Knowledge, and Greater Understanding.***

GLOSSARY OF TERMS

It is customary in academic writing to define the terms used in the document or book that is being written, and even though this manual departs somewhat from the rigors of academic writing, this is one feature that should not be excluded.

Science:

We define science as the compilation, cataloguing and analysis of evidence on any subject that is undertaken for study.

Sexuality:

The attraction one living being has for another regardless of gender.

SSA:

It is the automatic and involuntary attraction that people of the same gender have towards each other.

Religion:

The process through which humans seek to identify with the forces responsible for their existence. Its etymological roots are found in Latin. Re-ligare = Re-binding or Binding back

God:

The force responsible for the existence of the Universe.

Psychology:

The study of the psyche or the soul. A word that is also of Latin extraction. Psyche=soul Logos=study.

Racism:

It is the mistaken notion that some human beings are intrinsically superior to others.

Prejudice:

The act of judging someone or group negatively, prior to making the effort of getting to know them.

Fats:

A food product composed of long chains of carbon (C) atoms. Some carbon atoms are linked by single bonds (-C-C-) and others are linked by double bonds (-C=C-).

A **saturated fat:**

A type of fat in which the carbon atoms are all linked with single bonds.

Hydrogenation:

A process in which double bond atoms of fatty acids react with hydrogen to form single bonds. They are called saturated because the second bond is broken up and each half of the bond is attached to or saturated with a hydrogen atom. Saturated fats tend to be solids at room temperatures, while unsaturated fats tend to be liquid at room temperature

Trans fats:

The term trans fat is a highly complex industrial process in which hydrogen molecules are added to liquid vegetable oils to make them more solid. This process causes instability in the fat molecule making it less digestible to the human body. These fats raise the person's LDL levels (Low Density Lipids) better known as bad cholesterol, and they lower the body's good HDL (High Density Lipids) or good cholesterol, levels. If the hydrogen atom ends up on separate sides of the fatty acid chain it becomes a *trans*-fat.

"Trans" fatty acid

Cis fats:

When oils are hydrogenated, it is not possible to control where the hydrogen atoms land. If both hydrogen atoms end up on the same side of the structure, it is called a "Cis" fat.

"Cis" fatty acid

Free radicals:

Free radicals are atoms or groups of atoms with an impaired number of electrons. They occur through oxidation of the cell and are responsible for cancer, aging, and a variety of diseases. **Antioxidants:**

Antioxidants serve the purpose of combating the cellular damage caused by free radical. They are the body's most reliable defense system.

CELEBRATING HUMANITY

Notwithstanding the massive confusion that now engulfs the lives of nearly all members of the human specie; some time can be taken out to celebrate all that the specie has accomplished. We celebrate the science that gave us our much beloved cellular telephone, and cannot imagine our lives without it.

Those of us who have already achieved this level of thinking have decided against participating in any of the childish debates regarding god, belief or religion, and to relate to our fellow human beings solely on the bases of what we already agree on to possess in common. Some of the things we would hope to have in common are:

1- Creating a much better world for future generations, one that is based on reason, fairness and justice.

2- Empowering fair-minded people of all backgrounds to assert themselves despite the usual threats and intimidations people of good will invariably face when they manifest themselves.

3- Protecting healthiness of the environment 8 billion members of the species depend on for life and reproduction.

The Center for Intellectual Development, in association with Club Vizcaya International sponsors a series of concerts that celebrates humanity as it is, with all of its faults and its confusions. We've overcome Infanticide we've overcome widespread state sponsored slavery, we've overcome state condone rape and abuse of all type. This alone is a lot to celebrate, and that is the reason for our concerts and the people involved in promoting them.

The above examples are evidence that a better world has already been created, and it testifies to the fact that much more work is left to be done as we move to create a better world, one that is filled with meaningful dialogue fairness and justice.

We purposely leave out the term tolerance because it has been used a great deal as it relates to people with sexual natures other than that of heterosexual. The term is reserved to people of the LGBTQ communities, where the larger society has abrogated the right to tolerate these fellow humans. In celebrating humanity, the sacred rights of every human being to live happily is what we promote, and will battle anyone who aim at restricting those rights from any of our fellow humans because of the nature of their attraction.

Out of pure ignorance and religious nonsense, members of this community were made to suffer for centuries, especially after religions were invented. We now understand the complex nature of sexuality and sexual attraction, and we are immensely humbler as a result of this understanding.

Our concerts raise the funds needed for promoting a new system of building that create contemporary homes that are earthquake, fire and hurricane resistant, with accessible flat roof. At the moment getting this program off the ground constitute the greatest of all challenges, but it must be done.

If the overwhelming majority of the human race continues to toil in the darkest of ignorance, that is, ignorance as it relates to some of the most vital information about themselves ie; their own beginnings, their own humble origins, only one group can be blamed for it. That group undoubtedly is the academic class. It is important that we start from the top of that class which what has come to be known as the Ivy League Institutions. As research institutions these schools set the standards for what should be taught at all academic centers in the country and around the world. The tragedy of this is that these institutions, despite their stature and the glory given to them by the public at large, are in essence *Religious Institutions* lending support and endorsement to the childishness that is at the foundation of the *Abrahamic Faiths*. This religious trilogy created by what we define as the Abrahamic Faiths (Judaism, Christianity and Islam) is responsible for a great deal of human suffering. This is not to say that other invented religions like Buddhism, Hinduism, and Zoroastrianism among countless others, have not done their share of damage, the reality is that these three are the most influential in the times we live in. For all of their fame and glory, these Ivy League institutions are no more than giant churches masquerading as intellectual centers. The irony of this reality was underscored when the only African American president, a graduate of one of these institutions, declared his allegiance to Jesus Christ a mythical character not unlike that of Abraham or Moses. No one is sure who created Moses or how he was created, what we do know however is that Abraham before him was a figment of Moses' imagination. Not only that, the inventor of the Moses character makes him the writer of the *Pentateuch,* the first 5 books of the bible. In these 5 books the character of Abraham was created, and he was given powers that no other human had had up until that time, the power to see and speak to the god they'd recently invented. Approximately two thousand years later Jew residing in Greece, known as Hellenistic Jews, came up with the character of Jesus. With skillful manipulation the story took on a life of its own. The rest is history, and you challenge it at the risk of your own life, just as I am doing with these writings.

If all of the focus is on the evidence, it is infantile to spend time discussing mythology. Mythology is satisfying to some as they curl up into that world of imagination and comfort. But it does not work for science or the scientific mind. The scientific mind must have not just the facts but the actual evidence behind that so-called fact. At a time when we can work our way back to the birth of the Universe 14 billion years ago, who has the time to study the life, real or not, of men and women who lived and died just 2 or 3 thousand years ago.

Being human is all that matters today, and to be human is to embrace the totality of the experience of the modern members of this species which is only approximately 100.000 years old.

Born in Africa this species is still struggling to know itself, and the horrendous history of Africa does not make it any better. Everyone came out of Africa, waves of migration out of the continent produced multiple skin tones and physical characteristics, characteristics we've come to refer to as races. These races; another human invention not unlike religion and so many other concepts that have no bases in reality; are the source of much human suffering. Africa as the cradle of humankind and the birthplace of civilization is a foregone conclusion, at the highest academic circles this is no longer a debate. What is sad and pathetic is that they have not ventured to explain this to the ignorant masses. Perhaps it is because the masses are dugged so deep into their ignorance that words of wisdom on the part of the scientist would not get through. The time per hour is worth too much to be spent on the backward ignorant fools who without any factual information whatsoever convince themselves they're in the position to converse with, or worse yet, debate with scientists. Unfortunately, such is the case with the masses, few are willing to listen, and fewer yet are open-minded enough to allow evidentiary knowledge to take hold.

The fear of death is a uniquely human sentiment, one that came only with self-awareness. But scientists have been in a unique position to convince us that *life is energy, and energy never dies*. This is a simple and comforting response to our generalized fear of death

PSYCHOLOGY

I recently completed a book entitled Scientific Psychology, and this happen immediately before I came across the newest definition of psychology found in dictionaries or on the internet. They define it as: "The scientific study of the human mind and its functions, especially those affecting behavior in a given context."

The quackery and skullduggery that besotted this profession for over a century made it difficult for it to be respected as a solid scientific endeavor. But the discipline has grown considerably over the past two decades, and its new commitment to the evidence and what is real, is what makes of it a scientific endeavor. Brings then brings us into a much-needed definition for science. The newest and most contemporary definition for science is: "the intellectual and practical activity encompassing the systematic study of the structure and behavior of the physical and natural world through observation and experiment." That definition should be enough to give us pause because it focuses on the intellect. Unfortunately, that is not a popular word in our times. Today it is equated with boredom and being uppity. But there is no way around it. Our world is in desperate need of psychology, but the kind of psychologist who happen to be thinkers making use of the scientific approach. That in itself is difficult to find but it is a path down which we must go. Psychology is the only profession with the capacity to help human beings discover and understand themselves. Unfortunately, this discipline has not functioned with the confidence that is needed in order to play that crucial role.

That latest definition by which the discipline is described should be enough for the field of psychology to understand and carry out its mission, but the current obsession with breaking up the field into a million pieces with the introduction of a plethora of new psychological theories over the years, makes it nearly impossible for psychology to properly define itself and carry out its overall mission. If psychology is the scientific study of the human mind, nothing more needs to be said unless one wishes to confuse the issue. The fact is that science and psychology were never paired in the past. In fact, for most of its history psychology has been equated with quackery. If the intent is to carefully and scientifically study the human mind very little else needs to be said. The fact is that Psychology was born out of Philosophy, and philosophy is the first of all academic fields. "I think therefore I exist" that is the expression that defined philosophy since time immemorial. Psychology added feelings to the deep thinking demanded by philosophy.

It is the reality of our times that schools have not embrace the element of science in their teaching of psychology, and unless the student spend 10 years and a minimum of 110 credits after having achieved a master's degree, these schools do not feel their programs are good enough to support a student obtaining a state license for their practice. That intimidation that some schools experience, even when they are regionally accredited, has made it nearly impossible for more students to enter the field of psychology and this no doubt is a travesty. The world is in desperate need of psychologist and I know it. Social workers are attempting to fill that gap but these are entirely different animals. A social worker is not a psychologist, and to attempt to solve deep rooted psychological problems with a social work training or background reminds of the saying: "The road to *hell* is paved with a lot of good intentions." As a professional working in the field for 30 years I'm convinced that many of the unemployed holding a master's degree would do good to invest in a psychology PhD, but they should first join the efforts to modify the field and the people that are invested in destroying it i.e.; the American Psychological Association or APA. Unless they intend to be a national body certifying these PhD and licensing the to practice in this profession, that body has no reason to exist. The 50 state institutions in the United States that grant licensing are not made up of psychologist, and the few that there are have no connection to the scientific approach psychology must take if it is to survive as a science and as a profession. Children and adults the world over are in desperate need of the guidance and empowerment that only a psychologist grounded in scientific evidence can offer.

The reality of our times is that Psychology and the Psychotherapeutic process is not to be respected, hence the increase in teen suicide, suicide in general, drug addiction, and family disintegration, at all levels. The incompetence at the state level, are the principal culprits for this reality we face, not only in the United States but around the world. The American Psychological Association is greatly to be blamed for the conditions in the underserved dysfunctional population and the conditions in the mentally ill in that country for the conditions they've placed on the overall practice of psychology. One would think that with the hopelessness in the socioeconomic conditions overall, this body would be scrambling to create and license a lot more individual interested in tackling the problem but that is not the case. This group that up until a decade ago categorized Same Sex Attraction as a disease, it is surprising that they would remain as arrogant and as intransigent as they are at this stage, but they are, and that is what we are forced to contend with. Mothers with angry boys, and parents with unruly children have nowhere to turn to because the social workers currently encharged are in fact doing more harm than good. They have little and no background in in-depth psychology, and the psychotherapeutic process has an entirely different meaning for them than it does for psychologist. Many of them sit on various state psychology boards making the much-needed licensing process a nearly impossible task. It is sad and pathetic, but it is true. That happens to be the reality we're confronted with today.

For an illustration of the insanity as it relates to the profession of Psychology it must first be said that this is the only profession that requires the individual to be at the highest level academically; which is a PhD; before they can be considered a psychologist. Not only that, the profession is so steep in insanity that unless the degree is in clinical psychology it is hardly worth anything. The insanity of this is that regionally accredited graduate schools handing out PhDs in psychology have bought into the madness. Any student graduating from an APA accredited school is eligible for licensing in any state, yet the professors at these schools are so intimidated they've convinced their graduates that unless their concentration is in clinical psychology, they're not qualified for licensing. The laziness is astounding. All that is required is a few minutes of research to find out each state requirement for licensing, but even that appears to be too much for them. No doubt the licensing process is pathetic, but at least there is something in place for those who wish to pursue it. But the fact remains that the contemporary and fortified definition for psychology is all that we need. It is: *the scientific study of the human mind.* I use the term fortified because my 1971 College Edition of Webster's New World Dictionary already define psychology as a science. It defines it as: *the science dealing with the mind and with mental and emotional processes.* To be clear, if it is not *the scientific study of the human mind,* it needs not to be referred to as psychology. Anyone practicing this profession without the scientific approach is doing the world a disservice, and more than likely destroying their client. With so few individuals willing to take up the study of the mind we can understand the pathetic nature of our social world. People are powerless, desolate and in despair with no one possessing the capacity to reach out to them effectively. There plenty of well meaning, well intentioned persons, but again, *the road to hell is paved with a lot of good intentions.*

To be effective a psychologist must have a full and encompassing understanding of human history going back as far as the time long before there was life on this lonely planet. Advanced knowledge of every facet of their development and the exciting activities they've engaged in over the millennia is extremely important. Without that it becomes rather difficult to engage clients of different interest and different backgrounds coming to them with a multiplicity of problems. It is imperative to be in the position to assist client in stepping outside themselves and viewing the larger picture, a more global picture, a more Universal picture if you will. Only advance knowledge of everything and the wisdom and poise that ought to go with it, can bring comfort to those who are convinced their lives are meaningless.

SCIENTIFIC PSYCHOLOGY

Scientific Psychology sets as a premise for its intervention with human beings a working knowledge of the intricacies surrounding the origins of planet Earth, and the significance of this knowledge to the psychodynamics of the individual.

For nearly a century, scientists have argued that approximately 13 billion years ago an explosion took place that would culminate in the formation of stars and planets in infinite numbers. The first to formulate a theory with some resemblance to that was the English philosopher, theologian, and scientist Robert Grosseteste (1175-1253) who in his 1225 treatise *on the subject of light* explored the nature of matter and the cosmos. He described the birth of the universe in an explosion and the crystallization of matter to form stars and planets. He was referred to by many as the real founder of the tradition of scientific thought in medieval Oxford, and in some ways, of the modern English intellectual tradition.

Nearly 700 years later Monsignor Georges Lemaitre (1894-1966) a Belgian Catholic priest, astronomer and professor of physics at the Catholic University of Leuven arrived at a similar conclusion. He was the first to derive what is now known as Hubble's law and made the first estimation of what is now called the Hubble constant, which he published in 1927. Lemaitre is credited for having coined the phrase *Primeval Atom* and the *Cosmic Egg* to conceptualize the theory of the birth of the Universe.

The story here is that 2 highly positioned religious scientists, one English and the other Belgian were the authors of the big bang theory even though they were 700 years apart. This alone should end the non-existing rivalry between religion and science that we are eternally beset with.

Notwithstanding the fact that these theories were first presented by religious men for largely religious audiences, the theory itself is still difficult for most human minds to fathom. Endless questions about the source of that *Primordial Atom or that cosmic egg* that brought on the explosion would abound.

Accepting the notion of unfathomability that the scientist theory of the Big Bang brought us, contained within that story is one that is immensely more palatable to the average human mind. Those same scientists tell us that 8 billion years after that explosion (5 billion years ago), a cloud of cosmic dust and gases called a Nebula was formed on the outskirts of one of the billions of galaxies that makes up the Universe. One of those billions of galaxies is the one we belong to call the Milky Way. The nebula formed within its limits was replete with Hydrogen, Oxygen, Helium, Nitrogen and other elements. Gravity slowly drew the gas molecules together to form a spinning disk that sucked in even more hydrogen creating a great deal of pressure and heat at its center. These hydrogen molecules eventually fused to each other, igniting later and giving birth to a new star that is our sun. The debris present at the birth of our sun coalesced to form what we now know as planets, and the mass of the sun created a field of gravity that kept them swirling around it. One of those of course was to become our own planet Earth.

The planet remained a ball of fire for its first 2.5 billion years achieving temperatures that could sustain life only 2 billion years ago. Continuous bombardment of icy comets not only assisted in the cooling of the planet but assisted in the formation of its oceans.

In revealing to us that these chemical compounds: Carbon, Hydrogen, Oxygen, Nitrogen, Helium, and a few others are the building blocks of life and that they are found everywhere in the Universe, these scientists have really solved the major question facing all of humanity, *i.e.* What is life and how was it generated. With the information they provided we can deduce that life is an extraordinary combination of some of the most vital compounds found in the Universe. The fact that said experiment first took place in the waters of the newly formed oceans should not surprise us since water itself make up two of those chemical elements, Hydrogen and Oxygen. The combination of 2 molecules of Hydrogen and 1 molecule of Oxygen produces exactly that, water. There the first unicellular forms appeared more than four hundred million years ago, eventually growing lungs and legs to eventually step out of those waters to explore earth's vegetation. These first reptilian forms took on various sizes according to the conditions on the land.

The information these scientists revealed has also helped to confront our most dreadful fear, *death*. It begins with their definition of life as energy, adding further that in this Universe of ours, energy never dies that it simply transforms, mutating from one form into another. In other words, at the time our hearts stop permanently and we cease to breathe indefinitely, that energy is somewhere else participating in other functions.

On the question of life elsewhere in the Universe these scientists remind us that we are essentially star dust, that the same chemical materials found in our star is found in us, and that if that is so for our solar system why wouldn't it be for the other infinite number of solar systems that make up the Universe. The key element here is distance.

Our own star is 93 million miles away, but the second closest is Alfa Centaury at a distance of 4.37 light years away. To begin an understanding of this distance one must first understand that a light year is the distance light travels in the span of a year and the simple measurement of that distance begins with the understanding of the distance light travels in one second. The rest is simple multiplication as we do 186.000 x 60 x 60 x 24 x 365. Take that number and multiply it times 4.37 years and we get the distance that is to be covered to approach our nearest star. Traveling at the speed of light in little under 4 and a half years we'll make it to the vicinity of Alfa Centauri.

OUR EARTH'S PROTECTIVE LAYERS

The formation of the five invisible layers that protect the earth from the sun's ultraviolet rays and other threatening objects is another miracle we are yet to understand. Those five layers are:

1-The **troposphere** is the layer closest to Earth's surface. It is 4 to 12 miles thick and contains half of Earth's atmosphere.

2-The **stratosphere** is the second layer. It starts above the troposphere and ends about 31 miles (50 km) above ground. Ozone is abundant here and it heats the atmosphere while also absorbing harmful radiation from the sun. The air here is very dry, and it is about a thousand times thinner here than it is at sea level. It is the layer in which jet aircraft and weather balloons operate.

3- The **mesosphere** starts at 31 miles (50 km) and extends to 53 miles (85 km) high. The top of the mesosphere, called the mesopause, is the coldest part of Earth's atmosphere with temperatures averaging about minus 130 degrees F (minus 90 C). This layer is hard to study. Jets and balloons don't go high enough to study, and the orbit of satellites and space shuttles is just too high for them to take samples of that air. This is also where meteors burn up on their way to impact the earth.

4-The **thermosphere** extends from about 56 miles (90 km) to between 310 and 620 miles (500 and 1,000 km). Temperatures can get up to minus 2,700 degrees F (1,500 C) at this altitude. The thermosphere is considered part of Earth's atmosphere, but air density is so low that most of this layer is what is normally thought of as outer space. In fact, this is where the space shuttle flew in its time, and where the International Space Station is located at the moment. The aurora borealis and the aurora australis take place in this atmospheric layer. In this colorful phenomenon charged particles from space collide with atoms and molecules in the thermosphere, converting them into higher states of energy. The atoms shed this excess energy by emitting photons of light that produce the spectacle.

5-The **exosphere**, the highest layer, is extremely thin and is where the atmosphere merges into outer space. It is composed of very widely dispersed particles of hydrogen and helium.

COMMENTARY

The preceding is by far the most plausible explanation we have for the birth of this planet and the multiplicity of life form on its surface. It is not intended to be a lesson in astroscience but rather a presentation of some of the most basic information members are expected to manage in order to function effectively within the ranks of the Center for Intellectual Development. Our therapeutic interventions begin with this basic information, and it is designed not only to remove the notion of victimhood in our clients but also to empower them with a dose of the actual reality that brought about their existence.

This information levels the playing field as far as human relation is concerned. Where religion or religious belief sits within all of this is anyone's guess but it is not a part of our conversation. All thinking and all analysis must include this ultimate reality, without it the exercise is futile.

Today most treatment centers make reference to reality base therapy and brag about their ability to implement their version of this new approach. It is one of the few cases in which reality is accepted as a relative concept. It is almost as if they are redefining the term reality since in many cases these treatment centers subscribe to ideas principles and belief systems that are the furthest from reality. They treat patients with delusion and distorted thought processes while they themselves embrace the idea that a man is responsible for the creation of the world we know and everything in it. This delusional thinking has defined human affairs since time immemorial.

THE POWER OF KNOWLEDGE

Ten years ago, I published a book entitled The Human Experience. It was before I entered graduate school thus my writing skills and my patience as a scholar were not yet formulated. The re-publication of that book is still in the works but I've extracted this section for this booklet because of its relevance. The purpose of this is to keep our overall reality always in focus and above everything else. It reminds constantly that everything we do and everything that happens to us; regardless of its horror or intensity; is still a part of *The Human Experience.* It is also an integral part of the "intellectual regeneration" this organization is now determined to create. These facts regarding the sun and its influence on our planet are reproduced because they are a necessary component for sound thinking. We're confident that they will aide us in combating ignorance, and in taking this human race to another level in its relationship with the world and with each other. The following questionnaire and the list of facts that follows this representation is designed to initiate the process.

Questionnaire on the Fundamentals

1- Which is larger the sun or the earth is? _____

2- How much larger _____

3- Define Gravity _____

4- What is the relationship between the 2 bodies? _____

5- The sun's diameter is _____

6- The Sun's circumference is _____

7- The Earth's diameter is _____

8- The Sun's age is estimated at _____

9- The distance between the earth and the sun _____

10- The earth's circumference is _____

11- The speed of light is _____

12- The speed of sound is _____

13- The origins of the earth's oceans _____

FACTS TO LIVE BY

These facts are put together with one purpose only, and that is to combat **Ignorance** and **Powerlessness**, the two deadliest enemies known to mankind. We believe that it is rather unfortunate that facts such as these are only meaningful to a tiny few, the rest prefer to go about their lives enjoying the bliss of ignorance.

1-The first and most fundamental of all these facts is that the Sun is the source of all life on our planet.

2-The second of these is that the sun is 1.300.000 (one million three hundred thousand) time the size of the earth.

3-With its overwhelming superiority in size the sun exerts a gravitational grip over the earth keeping it beaming in its orbit at the speed of 66,000 miles per hour. To compare, the speed of a bullet is 5,600 miles hour.

4-The sun delivers five million (5.000.000) tons of pure energy each second. This energy is equivalent to four hundred million atom bombs, and the frequency is each second.

5-The sun's diameter is calculated at 865,000 miles, the Sun is composed of 80% Hydrogen, 18%Helium, and the remaining 2% is made up of Oxygen, Carbon, Nitrogen, Neon, Iron, Silicon, Magnesium, Sulfur, and other gases.

6-The Sun's circumference is 2.716, 100 miles

7-The Earth's diameter is calculated at 7,973 miles.

8-The Sun's age is estimated at 5.2 billion years.

9-The distance between the earth and the sun is calculated at 93.000.000 miles. *

10-The earth's circumference is 24,902 miles (40,075 km).

11-The speed of light is a constant 186,000 m/sec (300,000 km/sec).

12-The speed of sound is approximately 740 miles per hr. This speed is slightly greater in water, and even more so when traveling through iron.

13- Scientists can venture to tell us with supreme arrogance how the Universe got started 13 billion years ago with an unexplainable burst of energy but they have no idea where the earth got its waters from. After all, we were a flaming sphere for more than 2 billion years. Just ask them how the oceans were created and they all begin to stumble and mumble

The writer James Harvey Robinson in his book, *The Mind in The Making* stated:

> I venture to think that if certain seemingly indisputable historical facts were generally known and accepted, and permitted to play a daily part in our thought, the world would forthwith become a very different place from what it now is. We could then neither delude ourselves in the

simple-minded way we now do, nor could we take advantage of the primitive ignorance of others. All our discussions of social, industrial, and political reform would be raised to a higher plane of insight and fruitfulness. (p. 6).

Harvey is one of the few modern scholars that celebrated Francis Bacon and his accomplishments as a thinker, writer and intellectual. Bacon is also important because some research reveal that good many of the plays attributed to William Shakespeare were really written by him.

My own inclination towards Harvey Robinson should be obvious, he is one of the few writers that endlessly call for a higher level of thinking among humans, describing with clarity how beneficial this could be to all. This booklet and the organization it speaks for, has the same lofty goals.

These facts are fundamental and they constitute a shift in the way we view the world and our place in it. They constitute, if you will, a paradigm shift, and we like to refer to them as the building blocks of human intelligence. To ignore them is to embrace the confusion and hostility that presently defines human relations. They've been in place for billions of years despite our own ignorance of them, the western world became aware of them only a few hundred years ago with the revelation of Giordano Bruno, Nicholas Copernicus and Galileo Galilei. Embracing them changes the conversation entirely, and the subjects we argued over in the past such as: race, religion, social status and others, are no longer as relevant as we made them out to be. They represent what we like to refer to as the ultimate reality, clearing the mind of superstitions, and made up stories passed on through generations. Only a few short years ago, humans were convinced that they were at the center of the Universe and that the suns, as well as the other celestial bodies, are there for our pleasure and enjoyment.

Advances in science, scientific instrumentation and subsequently the advent of computers, allowed us to discover these as new facts, making them available for those passionate about knowledge and greater understanding. These facts are extremely helpful when practicing reality therapy because it inspires humility.

The speed with which the earth orbits the sun is both fascinating and humbling at the same time.

The total dependency of our earth on the sun and the overwhelming difference in size between the two, reminds even the most ravaged soul of the fact that their fate is equal to that of the six billion other lives on the planet. This new knowledge is empowering, because it liberates the human being from the silly fears of divine punishment or of going to hell. The figures involved in these facts have the potential to shock the senses and to bring us in contact with a new reality, one in which the individual is inspired to do more with his or her lives regardless of their present circumstances. In light of these new facts, all notions of inferiority and superiority disappear almost overnight.

It should be noted that the concept of reality therapy had been utilized for some time prior to this, however the introductions of these facts concerning our immediate and wider environment takes the whole thing to another level. With the mental stimuli that they provide, the individual is now prepared to think at a higher level, to be more creative and to be less of a victim in any situation, in short; to have the power to create their own sense of happiness.

It is humbling to think that were it not for our slow and painful growth through the ages human beings would be no more than a primate living on seeds, roots, fruits and uncooked flesh, wandering unclothed through the woods like a chimpanzee. This humility needs to be preserved if we are to achieve any significant growth as a species. Those struggling with an addiction are beginning to understand this notion, one that reminds them that without humility their chances of overcoming the crippling hold of the drug of choice is next to zero, leaving them at the mercy of an unending cycle of relapse and self-destructive behavior.

The origin and progress of humankind are still misunderstood and misconceived by the overwhelming majority of the population; one would hope that the clarification of these misconception would be high on the list of the educational system. It is imperative that we reconstruct our ideas of humans, and our capacities, freeing ourselves in the process from the persistent misapprehensions that weigh us down. The obstructionists, like the rationalizing theologians, and philosophers are all busy engaged in ratifying existing ignorant mistakes, and discouraging creative thought. In fact, few of us take the pains to study the origins of our cherished convictions; we have a natural repugnance to do so. On the contrary, we prefer to believe what we have been accustomed to accept as true, and when confronted, or when doubt is cast upon any of these assumptions we seek every manner of excuses for clinging to them.

34

This proves that our convictions on important matters are not the result of knowledge or critical thought, most of them are pure prejudices in the proper sense of the word. They are not our own ideas, but rather those of others who are no more informed or inspired than ourselves, having inherited them in the same careless and humiliating manner as we. They are; to borrow a quote from Harvey Robinson: *The whispering of the voices of the herd.*

By studying the manner in which human intelligence appears to have developed, we may eventually understand the perilous quandary in which mankind is now placed, and the way to escape that offers themselves. Ardrey (1972) states:

> We are a transitional species, nature's first, brief, local experiment with self-awareness, a head above the ancestral ape and a head below whatever must come next; we are evolutionary failures, trapped between earth and a glimpse of heaven, prevented by our sure capacity for self-delusion from achieving any triumph more noteworthy that our own sure self-destruction. (p.153).

We appear baffled by the senseless killings of our times while attempting the same old method for dealing with it i.e., the death penalty. But statistics have shown that the threat of the death penalty or its actual exercise, are yet to have an impact on the murder rate that consumes our modern society. We refuse to admit that there might be something in our savage nature that could surface at any time if the conditions are right. It can be triggered by a variety of factors, one of which is power over others, that is, the power to snuff out another life in an instant, in short, to play God. We're all a part of this mass insanity much of which stems from our inability to connect with our savage past. With the sense of power that comes with the weapon the out of control monstrosity we're witnessing at this point in our history should come as no surprise to us.

THE ENLIGHTENMENT ERA

In their book, *Telling the Truth About History,* Joyce Appleby, Lynn Hunt & Margaret Jacob introduce us to history through the prism of science. The first chapter of the book is titled: The Heroic Model of Science, and in it they walk us through the history of science from Copernicus (1473-1543) to the Industrial Revolution with an abundantly clear description of the mindset that accompanied each stage of this development. Their explanations are complex and intellectually challenging but within it we arrive at some truths regarding our present reality and how it came to be. In addition to Copernicus two other scientists played a significant role in breaking the grip the church had on the European society of the time. These were Giordano Bruno (1548-1600) and Galileo Galilei (1564-1642). Copernicus is credited with the heliocentric model for celestial bodies but fear of reprisals from the church prevented him from publishing his findings until he was near the end of his life. Bruno in turn was defiant, not only did he publish findings similar to that of Copernicus but he openly challenged the church, refusing to recant and was burned at the stakes. When Galileo published similar findings, he was denounced, imprisoned and even after recanting forced to spend his remaining years under house arrest. But the Genie was already out of the bottle; scientist and thinkers in Europe and elsewhere had learned of the heliocentric model and were challenging the church's authority on this subject. Many would argue that this sequence of events though spread out over more than a century, marked the beginning of the end for the dark ages. But the enlightenment age did not begin immediately after; strangely enough it was further research on the part of scientists in the protestant world that ushered in the enlightenment. It required thinkers and scientists the likes of Francis Bacon (1561-1626), Rene Descartes (1596-1650), Isaac Newton (1643-1727), Gottfried Leibniz (1646-1716), Emily Du Chatelet (1706-1749), Francois Marie Voltaire (1694-1778), Jean Jacque Rousseau (1712-1778), Denis Diderot (1713-1784), Mary Wollstonecraft (1759-1797), and Immanuel Kant (1724-1804) among others, to really usher in the new era, an era of science and reason. This new era would be marked by experiment, observation, mathematics and new forms of social communication. Established churches and religious dogmas were attacked: "as either deluded or upholders of backward-looking tyrannies, ignorance, prejudice, and superstition."[1] This enlightenment was spread over Western Europe and some parts of the thirteen colonies that were later to make up the United States of America. Appleby, Hunt and Jacob report that one of the founding fathers was also caught up in the frenzy of the enlightenment:" Late in the century Thomas Jefferson expressed faith in the link between science and progress by ordering a composite portrait of the life size bust of Bacon, Locke and Newton."[2] They also paint us a picture of elegant homes of entrepreneurs, merchants and aristocrats of that same period, adorned with miniature planetary systems with movable globes circling the sun in elliptical orbits, made by skilled workers in copper and wood. I have paraphrased here, but the contrast of this image with slaves of the period being indoctrinated with the most repressive brand of Christianity is striking.

---

[1] Appleby, et al. p 33
[2] Appleby, et al. p 25

At the heart of the enlightenment was Newtonian science, and it was also a key component of the Industrial Revolution, but at the height of the celebrations, the authors inform us that: "the same people who taught of themselves as enlightened, as teachers and appliers of Newtonian mechanics were often the profit seeking promoters of steam engines, canal companies, or factory style manufacturing."[3] The leaders of the enlightenment saw science as a means to improving the lives of humans but before long it was limited to serving the interest of greedy industrialists.

In an effort to arrive at the truth about history, science appears to play a unique role. It helps us to understand how we got to this stage in our development. Twisting the truth in this endeavor would defeat the purpose, so in some ironic way we are nudged into honesty. The need to know and understand forces us to tell the story of human development as it really occurred.

A commitment to truth, in the telling of history can carry with it enormous inspirational capabilities. To think that in less than three centuries we went from the dark ages to an enlightenment period is impressive, especially given the fact that this period was followed shortly thereafter by an explosion in technology. Some of this development was detrimental no doubt, but in all, great lessons have been learned. We now discuss facts rather than beliefs, and the wholesale slaughters of the recent past are no longer commonplace. We may teeter on the brink of total annihilation from time to time, but the painful lessons of the past are now burned into the hard drive that makes up our collective desire for self-preservation.

---

[3] Appleby, et al. p 23

Truth in history will remind us of our common ancestry as a specie, and though still resisted in many quarters, this factor is already playing a significant role in the healing process.

LESSONS IN COURAGE
SOCRATES AND GIORDANO BRUNO

In his (1964) publication entitled *Tales of Philosophy*, the Spanish physician Felix Marti Ibanez chronicled the life of some of Europe's most famous philosophers. For their bravery and commitment, we chose Bruno and Socrates as they best illustrate the point we have tried to make throughout this book. Both men paid the ultimate price for defending the truth; they paid with their lives. Socrates because he called on humans to reason and Bruno because he dared to say that the earth was round at a time when all Christians were convinced that it was flat.

SOCRATES

About Socrates, Ibañez (1964), tells us that a few thousand years ago, around 350 BC to be exact, a man by the name of Socrates walked the streets of Greece. His most famous quote was "life without philosophy is inconceivable". Socrates was born during the height of the Greek intellectual movement. They questioned everything, including religion and the democratic process, always teaching that truth was never objective or universal, but rather subjective and relative. History's most famous philosopher grew to manhood during Greece's wealthiest period, but he had no interest in riches. Seeing the multitude of articles displayed in the market, he once observed, "How many things there are that I do not want". He commonly wore the same rumpled tunic and walked the streets of Athens barefooted. He was trained as a youth in his father's profession of stone carver, won a reputation for bravery and superhuman stamina in the Peloponnesian War. He studied science as a youth, but abandoned it because it seemed a maze of mystery. There was a nobler effort to be made, said Socrates: the study of the mind of man and the quality of his life, thus, as Cicero commented, Socrates brought philosophy down from heaven to earth. Some called him a sophist, but he differed from the sophists by disliking rhetoric and by desiring to strengthen morality instead of weakening it. He consistently avowed that he taught no more than the art of examining ideas when the oracle Delphi was reported to have pronounced him the wisest of men, he protested that he did not possess wisdom but only sought it. Socrates did not preach a return to old dogmatic beliefs but asked such searching questions as: is a natural ethic possible? Can morality be independent of religion? He propounded that men were evil because they were ignorant and thoughtless: if they could see the good, they would choose it; and right thinking would inevitably lead to right conduct. Said he: *Virtue is Knowledge.* If man is the measure of all things, the art of living must begin with the individual and he must question old beliefs, traditions, and creeds, taking nothing for granted in his search for truth. The method of the Socratic dialect was to seek knowledge by question and answer; this probing was intended to expose the shallowness or error of definitions, to expose harmful and false opinions. "Know thyself "was the dictum by which he stung his fellowmen to unflinching self-examination, and when his opponents spoke glibly of such abstractions as courage, fear, honor, he defied them to define their term. He was in high favor as long as Athens was prosperous and powerful, but when that proud city finally fell to Sparta, many leading citizens were scapegoats of the national disgrace and Socrates was no exception. Charges were trumped up that Socrates was setting up gods and corrupting Athenian youth. He was brought to trial found guilty and condemned to end his life by drinking hemlock. To his sorrowing friends he said. "Be of good cheer and say that you are burying my body only". Then he serenely drained the cup of poison that gave him death and immortality. His influence on the development of human thought was enormous, and his emphasis on placing conscience above the law later became a cardinal tenet of Christianity. His insistence on the clear definition of general concepts, his approach to inductive reasoning, and his faith in an intellectual and moral order remain a force to this day (p.28). He was the first great European whose death in 339 BC marks the birth of Europe because he was the first martyr for European faith. Socrates was the first European to use intellect as an instrument for obtaining knowledge of the Universe. In his belief in the divine nature of the human soul, he was a precursor of Christ. He chose truth as a guide for humans (p.21). Even with the mind of Socrates and many others whose stories never really gained notoriety Europe continued to grow, albeit ever so slowly. Italy became the seat of power through Rome and its military conquests, after all the Roman Empire lasted a thousand years. But as it shrank in size and in might the Church became the seat of power and they could command what was left of the army to do their biddings in drowning the voices of those who wished to think, to study and to explore. It is in this environment that scientist such as Copernicus, Galileo, and Giordano Bruno were compelled to do their work. The church in its backwardness would consider these men to be heretic and would seek to punish them.

GIORDANO BRUNO

One individual that attempted to shift that paradigm was Giordano Bruno (1548-1600). Bruno was born 5 years after the death of the great Nicolas Copernicus, and he was considered to be a contemporary of Galileo who the story goes was critical of him, but 13 years his junior Bruno had little time for Galileo, he was more concerned with staying alive as his ideas had already enraged the church. The story has it that this Dominican friar born in Italy in 1548 and christened with the name Fillipo Bruno was a philosopher, a mathematician, poet and cosmologist, among other things. Bruno's interests in cosmology lead him to support the heliocentric ideas of the Polish astronomer Nicolas Copernicus (1473-1543). At the time, Copernicus was one of the few Europeans to embrace the notion that it is not the Sun that revolves around the Earth, as it was taught by the clergy at the time, but rather the Earth that revolves around this celestial body in its yearly journey. To go against the sacred teachings of the church was considered heresy, and at the time heresy was punishable by death. The power of the clergy instilled fear in Copernicus, and because of that he refused to publish his scientific findings.

History tells us that 21 years after the death of Copernicus one of his most ardent followers namely, Galileo Galilei (1564- 1642) was to come into existence. Free from the fears that paralyzed his hero Nicolas Copernicus, Galileo set out to prove those theories that the earth is not at the center of the world, and that rather than the sun revolving around our planet, it was us that made the yearly journey around the sun. For his boldness and his conviction, Galileo paid a heavy price. He was forced to recant and renounce his theory, and even after doing so he was placed under house arrest for the remainder of his life.

The dispute between Bruno and Galileo amounted to very little, they both embraced Copernicus' ideas, and they both suffered for it, what appears to have bothered Galileo about Bruno was his open and confrontational attitude towards the clergy. They were the power, and to confront them spelled death not only for the individual that does so but for anyone else who shares his ideas. Consequently, after they murdered his colleague Bruno, tied at the stakes in living flames, he Galileo bore the brunt of the church's ire, and only a recant of his own ideas could save him from certain death at the hands of the church.

There is no question that the hero of heroes in this story is Giordano Bruno for even though he too recanted a few times, and went in hiding for long periods, he eventually stood to the vicious power of the church that murdered anyone who dared to hold contrary views to theirs.

Ibanez (1964) also wrote extensively about Bruno and in his narrative, he stated that in the XIV century as Europe attempted to shake the yoke of its medieval past, there was an awakening to science and the infinity of space commanded the attention of most thinkers. Among these thinkers there was Nicholas Copernicus born in Poland in 1473. His fascination with astronomy had him to construct a crude observatory, and from his findings elaborated the theory that the earth and all other planets evolved around the sun. Fear of the church's censure made Copernicus delay the publication of his theory until 1543 when he lay dying.

Bruno was born five (5) years after in 1548. He inherited the same interest in astronomy, as did Copernicus. He also studied religion, becoming an outspoken Dominican Friar whose free thinking and fearlessness forced him to flee Venice after charges of heresy were brought against him. Two years later, in Geneva he embraced Calvinism, was imprisoned briefly for publishing an attack on a professor, and after his release taught philosophy at Toulouse. For much of his life Bruno was a lonely wandering philosopher, bombastic, arrogant, with a remarkable ability to outrage authority. He debated with Oxford professors over Copernicus, embraced a pantheistic doctrine that conjoined God and nature as the active and passive element of reality (p.112).

A desire to make peace with the church drew him to Venice, but he was quickly arrested. It was now 1593 and the inquisition was in full swing so they kept him imprisoned for eight years in an attempt to have him change his beliefs. Under questioning, he stated: "I hold the Universe to be infinite, as being the effect of divine power and goodness. Hence, I have declared infinite worlds to exist beside this our earth. It would not be worthy of God to manifest Himself in less than an infinite Universe." Refusing to recant, he told his judges; "You are more afraid of this than I am," and was burned at the stake in1600.

GALILEO GALILEI

Perhaps the saddest of all these stories was that of Galileo Galilei (1564-1642), son of a distinguished philosopher in Roman Italy. Destined at first to become a physician, he became far more interested in Euclid and Archimedes and devoted his time to experiments in physics, including the famous demonstration from the leaning tower of Pisa on the laws of falling bodies. He was the first to build an optic telescope and find, as he said, "with incredible delight" that he could distinguish the mountains on the moon and spots on the sun, discover stars in the Pleiades, observe satellites rotating around Jupiter and obtain glimmering of Saturn and its mysterious rings. His discoveries, which confirmed his belief in the Copernican system, raised a storm of protest: the defenders of the old Ptolemaic system accused the new cosmology of contradicting parts of the Scriptures, and in 1616 the theologians of the Holy Office decreed the heliocentric theory to be heretic. Galileo was warned not to teach the forbidden doctrine, but in 1632 he published his great work in elegant Italian, "Dialogues on the Two Principal Systems of the World" which unflinchingly supported the Copernican system with many of his own observations. He was summoned before the Inquisition in Rome, made to recant, and sentenced to retire in strict seclusion. In 1637 he made his last astronomic discovery, and a few months later became blind. He sadly wrote: "These heavens, this earth, this universe, which by wonderful observation I had enlarged thousand times are henceforth dwindled into the narrow space which I occupy." Galileo died of a fever in1642, the very year Isaac Newton was born (p.lll). The story of the European's quest for knowledge unit the painful birth of western civilization is not at all the history of mankind, because many civilizations preceded theirs and, in most instances, they were aware that their earth was not the center of the Universe. It took the Europeans several centuries to overcome that thought. In these stalwart bastions of religious darkness, efforts continue to limit human thought and to promote fear and ignorance as a way of life. In order to achieve happiness, you must place yourself above the mundane. You must have the ability of viewing the world and all its nuances from a vantage point that is far above.

ÉMILIE DU CHÂTELET (1706 -1749)

Émilie du Châtelet was a French natural philosopher, mathematician, physicist, and author during the early 1730s until her untimely death due to childbirth in 1749. du Châtelet's father Louis-Nicolas, recognizing her early brilliance, arranged for Fontenelle to visit and talk about astronomy with her when she was 10 years old Her most recognized achievement is her translation of and commentary on Isaac Newton's book *Principia* containing basic laws of physics. The translation, published posthumously in 1759, is still considered the standard French translation today. Her commentary includes a profound contribution to Newtonian mechanics—the postulate of an additional conservation law for total energy, of which kinetic energy of motion is one element.

Her book entitled the *Foundations of Physics*, circulated widely, generated heated debates, and was republished and translated into several other languages within two years of its original publication. She participated in the famous *vis viva* debate, concerning the best way to measure the force of a body and the best means of thinking about conservation principles. Posthumously, her ideas were heavily represented in the most famous text of the French Enlightenment, the *Encyclopédie* of Denis Diderot and Jean le Rond D'Alembert, first published shortly after Du Châtelet's death. Numerous biographies, books and plays have been written about her life and work in the two centuries since her death. In the early 21st century, her life and ideas have generated renewed interest.

Du Châtelet's mother, Gabrielle-Anne de Froulay, was brought up in a convent, at the time the predominant educational institution available to French girls and women.[5] While some sources believe her mother did not approve of her intelligent daughter, or of her husband's encouragement of Émilie's intellectual curiosity,[5] there are also other indications that her mother not only approved of Du Châtelet's early education, but actually encouraged her to vigorously question stated fact.[6]

Du Châtelet and Voltaire may have met in her childhood at one of her father's *salons*; Voltaire himself dates their meeting to 1729, when he returned from his exile in London. However, their friendship began in earnest in May 1733, upon her re-entering society after the birth of her third child.

Du Châtelet invited Voltaire to live in her country house at Cirey-sur-Blaise in Haute-Marne, northeastern France, and he became her long-time companion (under the eyes of her tolerant husband). There she studied physics and mathematics and published scientific articles and translations. To judge from Voltaire's letters to friends and their commentaries on each other's work, they lived together with great mutual liking and respect. Sharing a passion for science, Voltaire and Du Châtelet collaborated scientifically. They set up a laboratory in Du Châtelet's home. In a healthy competition, they both entered the 1738 Paris Academy prize contest on the nature of fire, since Du Châtelet disagreed with Voltaire's essay. Although neither of them won, both essays received honorable mention and were published.[12] She thus became the first woman to have a scientific paper published by the Academy.

Her book *Institutions de Physique* ("Lessons in Physics") appeared in 1740; it was presented as a review of new ideas in science and philosophy to be studied by her thirteen-year-old son, but it incorporated and sought to reconcile complex ideas from the leading thinkers of the time. The book and ensuing debate contributed to making her a member of the Academy of Sciences of the Institute of Bologna in 1746.

Du Châtelet's contribution was the hypothesis of the conservation of total energy, as distinct from momentum. Inspired by the theories of Gottfried Leibniz, she repeated and publicized an experiment originally devised by Willem's Gravesande in which balls were dropped from different heights into a sheet of soft clay. Each ball's kinetic energy - as indicated by the quantity of material displaced - was shown to be proportional to the square of the velocity. The deformation of the clay was found to be directly proportional to the height the balls were dropped from, equal to the initial potential energy. Earlier workers, such as Newton and Voltaire, had all believed that "energy" (so far as they understood the concept at all) was indistinct from momentum and therefore proportional to velocity. Energy must always have the same dimensions in any form, which is necessary to be able to relate it in different forms (kinetic, potential, heat...). Newton's work assumed the exact conservation of only mechanical momentum. A broad range of mechanical problems are soluble only if energy conservation is included. The collision and scattering of two point masses is one of them. Leonhard Euler and Joseph-Louis Lagrange established a more formal framework for mechanics using the results from du Chatelet.

She is but one of the many women scientists this species has been fortunate enough to have had over the centuries some equal in brilliance and others whose brilliance may have surpassed hers.

One shining example is that of the female mathematicians that made the United States space mission possible. They were referred to as human computers because of the massive figures they were able compute, and because of the complex data they were able to analyze before the advent of computers. Their story is told by Margot Lee Shetterly and played out in a movie by the same name. The book chronicles the lives of Katherine Johnson, Dorothy Vaughan, and Mary Jackson, three mathematicians who overcame discrimination, as women and as African Americans, while working at the National Aeronautics and Space Administration (NASA) during the Space Race. For the first years of their careers, the workplace was segregated and women were definitely kept in the background as Human computers. Their calculations were a determining force in America's greatest achievements in space.

Before John Glenn orbited the earth, or Neil Armstrong walked on the moon, this group of dedicated female mathematicians used pencils, slide rules and adding machines to calculate the numbers that would launch rockets, and astronauts, into space. They were exceptionally talented African American women, some of the brightest minds of their generation. Originally relegated to teaching math in the South's segregated public schools, they were called into service during the labor shortages of World War II, when America's aeronautics industry was in dire need of anyone who had the right stuff.

Suddenly, these overlooked math whizzes had a shot at jobs worthy of their skills, and they answered Uncle Sam's call, moving to Hampton, Virginia and the fascinating, high-energy world of the Langley Memorial Aeronautical Laboratory. Even as Virginia's Jim Crow laws required them to be segregated from their white counterparts, the women of Langley's all-black "West Computing" group helped America achieve one of the things it desired most: a decisive victory over the Soviet Union in the Cold War, and complete domination of the heavens. Starting in World War II and moving through to the Cold War, the Civil Rights Movement and the Space Race, Dorothy Vaughan, Mary Jackson, Katherine Johnson and Christine Darden, participated in some of NASA's greatest successes while they used their intellect to change their own lives, and their country's future.

THE REAL STORY BEHIND HIDDEN FIGURES

A little over 200 years after Duchatelet's death a group of black scientists made history as they played a pivotal role in the North American Program. Haas (2016) reports that as America stood on the brink of a Second World War, there was also a strong appetite for the idea of conquering space. But this appetite could not be satisfied without the calculations and analysis of advanced mathematicians.

Interestingly enough it was women who answered that call as they were recruited by the Langley Memorial Aeronautical Laboratory in 1935 to shoulder the burden of these calculations long before computers or calculators were made available. Even though it was a government program conducting war plans the Langley Memorial Aeronautical Laboratory was a segregated institution so Black scientists and mathematicians were not allowed to participate.

Franklin Delano Roosevelt was president at the time and the civil rights leader A Phillip Randolph approach him with the unfairness of this social reality. Randolph threatened a march on Washington, D.C., to draw attention to these and other forms of racial discrimination. With the threat of 100,000 people swarming to the Capitol, President Franklin D. Roosevelt issued Executive Order 8802, preventing racial discrimination in hiring for federal and war-related work. At the time, a company called West Computers located in Langley Virginia had been active in conducting the equations that described the functioning of the newly produced airplanes, so when Roosevelt signed his executive order an army of sharp minded Black women were hired by the company to assist in doing the complex calculations that would make these machines safer faster and sleeker. Unrest was brewing in Germany and analysts could tell that another war was afoot that could again require the participation of the United States.

These African American female calculators became a computing pool for specific projects, some like Christine Darden worked to advance supersonic flight, Katherine Johnson calculated the trajectories, launch windows, and emergency back-up return paths for many flights from Project Mercury, including the early NASA missions of John Glenn and Alan Shepard. She was instrumental in the 1969 Apollo 11 flight to the Moon, and was active through the Space Shuttle program.

THE POWER OF WOMEN

After examples like those of Emilie Duchatelet and the sharp minds that made the space program possible, it would not be fair to complete this small pamphlet without an attempt to enhance the value of women, not in the romantic sense as it usually is but as representing more than half of humanity. Instead of regretting the fact that she has never been given the respect and praise she deserves, the language here represents the search for the most effective way for women to begin to exercise their power on this planet. The latent power of women is extraordinary, and it is only a matter of time before that power is demonstrated latent because unless it possesses and believes in the scientific information offered here, it will continue under the psychological influence of a structure invented by men, the very men who have declared her as an inferior being. For reasons that women do not control in almost every country in the world they have fallen into the male version of the creation of this world. Her identity has been defined by the masculine element, and although she recognizes it, it is quite possible that the complexity of the subject still escapes her. Unless women are willing to reject the invented ideas of religion that men have imposed on them over the millennium, any effort to be liberated will be futile.

THE EVOLUTIONARY PROCESS

Despite the fact that this book is more about the origins of planet earth, life on it, and the impact this has on psychology, the species that those with the ability to read these pages belong to is of great consequence. It goes without saying by now that race, ethnicity and culture are all of little or no consequence when it comes to the origins of our planet. The fact that the species we call humans originated on the continent of Africa is of great consequence.

Ardrey (1963) argues that the roots of human ancestry is steeped in violence, and he does so by quoting the South African anthropologist Raymond Dart's 1953 paper entitled *The Predatory Transition from Ape to Man* in making his case. Dart's study leads him to the conclusion that man's ancestors were killer apes and that their weapons of choice in those early days had been the antelope humorous bone. Ardrey (1963) stated that what Dart put forward in his piece was the simple thesis that Man had emerged from the anthropoid background for one reason only, and that is because he was a killer. Ardrey goes on to quote Dart as saying:

> Long ago, perhaps many millions of years ago, a line of killer apes branched off from the non-aggressive primate background. For reasons of environmental necessity, the line adopted the predatory way. For reasons of predatory necessity, the line advanced. We learned to stand erect in the first place as a necessity of hunting life. We learned to run in our pursuit of game across the yellowing African savannah. Our hands freed from the mauling and the hauling, we had no further use for a snout; and so it retreated.

And lacking fighting teeth and claws, we took recourse by necessity to the weapon (p. 29).

Through Ardrey (1963), Raymond Dart goes on to present his case that this weapon could be a rock, a stick or a heavy bone and that in either of these cases it would have meant the margin of survival for our ancestral killer, but he added that the use of the weapon represented new and multiplying demands on the nervous system for the coordination of muscle, touch and sight. This combination of factors, argued Dart, contributed to the creation of a larger brain, a necessary requirement for modern man. Since that period, argued Dart, man has become a slave to the weapon, in other words, the weapon had fathered man (p.29). Dart's theory reflects significantly on the horrors on Nazi Germany despite the fact that it avoids the concept of the mistreatment of children as a contributing factor to the horrors we've all become so familiar with.

Ardrey (1963) does not answer the question of whether violence is imprinted in the species' DNA but the hypothesis regarding the weapon comes close to making that particular point. If Dart is right and violence has been with the species since its genesis, then it may well be that only through emotional and intellectual growth taking place at a massive level, that we will be able to overcome our violent tendencies.

UNDERSTANDING SEXUALITY

In the glossary of terms Same Sex Attraction was in the introduced as a way of joining the dialogue on the subject related to this naturally occurring phenomenon.

As we speak of the power of knowledge it is imperative that we also include the destructive nature of ignorance. Throughout history little boys and little girls that find themselves attracted naturally to the same sex have been condemned verbally, castigated physically and have been ostracized for something beyond their control. In many areas of the United States; a supposedly developed country; and in many other countries around the world, the ignorance that produces this level of mean-spiritedness goes on unabated.

At the Center for Intellectual Development and its sister organization Club Vizcaya International, it is not the acceptance or rejection of LGBTQ+ community at is discussed but rather the prevailing ignorance of such a large portion of the human family. Rather than discussing the acceptance of someone's nature, our mission is to expose the horrifying nature of the ignorance that allows any human being to be convinced that he or she is superior to another solely because of their sexual nature.

In their suffering, what these young men and women did was reveal to us the massiveness of our own ignorance. It is for that matter that no one holding discriminatory views on another person's sexual nature; will be allowed to operate within either of these organizational structures.

Our mission is to combat ignorance in all of its ugly manifestations, and potential members are welcome into the fold of this organization once they rid themselves of these horrifying and demeaning thoughts.

Part of the reeducation of these potential members includes the understanding that same sex attraction is also a natural occurrence among several other species within the animal kingdom. That said, religious scriptures of any kind are disallowed within this organization, and discussions based on teachings that are born out of ignorance are not accepted. Members are free to practice the religion of their choice, but under no circumstances will the ignorance and prejudices fomented within these religious structures be tolerated here.

ALAN TURING (1912-1954)
THE GAY MAN WHO SAVED THE WORLD

Alan Turing was an English computer scientist, mathematician, logician, cryptanalyst, philosopher and theoretical biologist. He was highly influential in the development of theoretical computer science, providing a formalization of the concepts of algorithm and computation having introduced the world to the Turing machine, at the time a model of a general purpose computer. Turing is widely considered to be the father of theoretical computer science and artificial intelligence. During the Second World War, he worked for the Government Code and Cypher School (GC&CS) at Bletchley Park, Britain's code breaking centre that produced Ultra intelligence. For a time, he led Hut 8, the section which was responsible for German naval cryptanalysis. Here he devised a number of techniques for speeding the breaking of German *ciphers* including improvements to the pre-war Polish *Bombe,* an electromechanical machine that could find settings for the *Enigma* machine. He played a pivotal role in cracking intercepted coded messages that enabled the Allies to defeat the Nazis in many crucial engagements, including the Battle of the Atlantic, and in so doing helped win the war Cooper (2013).

His earlier theoretical concept of a universal Turing machine had been a fundamental influence on the Manchester computer project from the beginning. After Turing's arrival at Manchester, his main contributions to the computer's development were to design an input-output system—using Bletchley Park technology—and to design its programming system. He also wrote the first-ever programming manual, and his programming system was used in the Ferranti Mark I, the first marketable electronic digital computer (1951).

Turing was a founding father of artificial intelligence and of modern cognitive science, and he was a leading early exponent of the hypothesis that the human brain is in large part a digital computing machine. He theorized that the cortex at birth is an "unorganized machine" that through "training" becomes organized "into a universal machine or something like it." Turing proposed what subsequently became known as the Turing test as a criterion for whether an artificial computer is thinking (1950). For his work he elected a fellow of the Royal Society of London in March 1951 one of the highest of honors.

After the war, Turing worked at the National Physical Laboratory, where he designed the ACE, among the first designs for a stored-program computer. In 1948 Turing joined Max Newman's Computing Machine Laboratory at the Victoria University of Manchester, where he helped develop the Manchester computers and became interested in mathematical biology.

Turing was arrested prosecuted in 1952 for homosexual acts, considered criminal in England at the time. He accepted chemical castration treatment, as an alternative to prison but with a criminal record he would never again be allowed to work for Government Communications Headquarters (GCHQ), the British government's postwar code-breaking centre. In the midst of this groundbreaking work, Turing was discovered dead in his bed, poisoned by cyanide. The official verdict was suicide, but no motive was established at the 1954 inquest. His death is often attributed to the hormone "treatment" he received at the hands of the authorities following his trial for being gay. Yet he died more than a year after the hormone doses had ended, and, in any case, the resilient Turing had borne that cruel treatment with what his close friend Peter Hilton called "amused fortitude." Also, to judge by the records of the inquest, no evidence at all was presented to indicate that Turing intended to take his own life, nor that the balance of his mind was disturbed (as the coroner claimed). In fact, his mental state appears to have been unremarkable at the time. Although suicide cannot be ruled out, it is also possible that his death was simply an accident, the result of his inhaling cyanide fumes from an experiment in the tiny laboratory adjoining his bedroom. Nor can murder by the secret services be entirely ruled out, given that Turing knew so much about cryptanalysis at a time when homosexuals were regarded as threats to national security (Copeland 2005).

By the early 21st century Turing's prosecution for being gay had become infamous. In 2009 British Prime Minister Gordon Brown, speaking on behalf of the British government, publicly apologized for Turing's "utterly unfair" treatment. Four years later Queen Elizabeth II granted Turing a royal pardon. (Copeland 2005).

A STUDY OF CHILDHOOD
JEAN JACQUE ROUSSEAU

In this chapter I will discuss some of the concepts and theories surrounding the issue of child development, particularly those concepts that focus on the role of experience in learning, in childhood. The theorists that are profiled in this essay are: John Dewey, Maria Montessori, Lev Vygotsky, Jean Piaget, and Erik Erikson. The views of these noted theorists will be compared and contrasted with that of one who preceded them all by more than a century, Jean Jacques Rousseau (1712-1782). I will attempt to explore whether any of these theorists derived any of their ideas from Rousseau or if they departed completely from Rousseau's thoughts.

Rousseau was chosen for a number of reasons, but chief among them is his unwavering commitment to the safety and security of children; at least as manifested through his writings. The controversies, and contradictions surrounding Rousseau's life also makes him a compelling figure for study and analysis for many scholars wishing to develop some understanding of or learn from the life of this enigmatic historic figure. Perhaps the most noted contradictions in Rousseau's life is his abandonment of his five children; none of multiple birth; to a foundling institution never to see them again. Scholars must make extraordinary to understand or explain away these and other controversial aspects of Rousseau's life.

Rousseau is considered by many to be one of the first successful novelists, having rendered his opinions on children through two famous novels: *Emile* and *The New Heloise*. Rousseau presents the ideas for his character Emile through a series of five books which have come to represent five of the stages of his character's life, and although, like Rousseau, the other thinkers and theorists dealt with in this report, all subscribe to the concept of stages in a child's development, this report will focus on the complexities involved in the adult child relationship described by Rousseau.

At a time when the concept of the child was not yet defined, Rousseau endowed the child, and particularly the infant, with almost magical powers in their ability to command the attention of those around them, and the enormous care that had to be invested in securing its safety. We find in Rousseau (1979):

> At birth the child cries; his earliest infancy is spent in crying. Sometimes he is tossed, he is petted, to appease him; sometimes he is threatened, beaten, to make him keep quiet. We either do as he pleases or else we exact from him what pleases us; we either submit to his whims, or make him submit to ours. There is no middle course, he must either give or receive orders. Thus, his first ideas are of absolute rule, and of slavery. Before he knows how to speak he commands; before he is able to act, he obeys; and sometimes he is punished before he knows what his faults are, or rather; before he is capable of committing them. Thus, do we pour into his young heart the passions that are later imputed to nature; and after having taken pains to make him wicked, we complain of finding him wicked (p.21).

Profound in its scope this analysis gives us as clear a picture of any other of the complexities embodied in the infant child and the struggles of adults to relate adequately to this creature they themselves grew out of a few years earlier. This highly analytical and deeply psychological statement was made more than a century before the birth of the discipline we have come to know as psychology.

The dilemma that lies in reducing the powers the child holds, contrasted with the willingness or lack thereof on the part of adults to set this child free and allow it to grow unencumbered is one that will be dealt with in this book. In short, more liberty and less power, since the child neither requests nor has any need for the powers we attribute to it and it certainly is not happy when the other side of that power is applied to him or her in the form of punishment or other corrective measures.

It is this dilemma that will be explored here as we assess the approach of other thinkers and theorists to this most complex of subjects.

With some degree of accuracy, it could be said that this unresolved dilemma lies at the heart of much of the mistreatment meted out to children for all of recorded history, since in all that time as in the present, the specie struggled with devising efficient methods for raising children. That said, it can also be argued that Rousseau was one of the first to advocate for the safety and security of children, doing so without preceding theories or models that we know of. Thus, the striking of a balance between the freedom which the child yearns, and those magical powers attributed to it that it never asked for, will be explored in this book along with the opinions of others theorists and thinkers.

Mooney (2000) tells us that Dewey was born in Burlington Vermont on October 20th, 1859 into a farming family. He studied philosophy at the University of Vermont from where he graduated in 1879 with a degree in philosophy. Dewey went on to do graduate work at the John Hopkins University where he obtained a PhD in philosophy in 1884. After graduating, he accepted a teaching position at the University of Michigan. In 1894 he was offered a position at the University of Chicago that allowed him the opportunity to combine his teaching of philosophy, with two other disciplines: Psychology and Educational Theory. Mooney (2000) tells us that within 2 years he had established the famous laboratory school that attracted attention around the world. Dewey's Laboratory School established the University of Chicago as the center of thought on progressive education, the movement toward more democratic and child-centered education. Mooney (2000), tells us that Dewey's position at the head of the lab school was relatively short-lived but created, in a few years, a wealth of educational research and theory that continues do drive many of our best practices today (p.2).

In Dewey (1916) we find a rather sober reference to youth and our responsibility towards them when he argues that in directing the activities of the young, society determines its own future. The nature of the child will largely turn upon the direction children's activities were given at an earlier period. Dewey describes this cumulative of activities towards becoming a better person, as growth. (p.49) Dewey focuses on the powers children possess for enlisting the cooperative attention of others, which he claims is another way of saying that others are marvelously attentive to the needs of children. That they are egotistical and self-centered before adolescence is not lost on Dewey, but it is his conviction that even in their egotistical self-centeredness there are moments when they are able to capture everyone's heart. Stable adults recognize that this power the child possess for garnering their attention is only temporary, and even their egotistical self-centered behavior becomes more tolerable because again it is only temporary. The travesty Dewy argues, are those adults too absorbed in their own affairs to take any real interest in children's affairs (p.52).

Mooney (2000) presents John Dewey's *Pedagogue Creed,* published in 1897. In its first and second articles Dewey makes reference to the power of children. He says "True education comes through the stimulation of the child's powers ... The child's own instincts and powers furnish the material and give the starting point for all education" (p.4). Here Dewey speaks, not so much of the powers of the infant but of with the potential powers of the child in the learning process.

In what might be one of John Dewey's most significant statements we have in Dewey (1916), "From a social standpoint, dependence denotes a power rather than a weakness" (p. 52), for it involves interdependence, and a child's gift for social interaction, coupled with its dependency all tend to facilitate social responsiveness and social interaction. Rather than being in awe with a child and its immaturity, surreptitiously granting the child all of that unrequested power, Dewey (1916) argues that respect for immaturity can actually work as a perfect antidote for all that power we give to the child and the resentment that naturally accompanies it, for it is in the resentment that the attitude towards the child changes thus placing them in danger. Evoking the words of Emerson, Dewey (1916) implores us to respect the child, its space its right to solitude, for in the end respect for the child invariably translates to respect for self (p.62). But returning to the subject of freedom Dewey (1916) reminds us that the important thing to bear in mind is that it involves a mental attitude rather than an external constraint of movement, but also that this quality or state of mind cannot be achieved without a freedom of movement that allows one to explore, to experiment, and to apply all that has been learned. Applied to the child, this concept may allow us to envision a healthier society, one with less cases of depression, or aggression.

Mooney (2000) then introduces us to Maria Montessori, the Italian physician, and education reformer. Although she argued for educational structure, Montessori shared Rousseau's views that freedom and play constitute essential components in a child's development, and that it is in these unstructured activities that much of their learning takes place. That it is in this natural unencumbered environment that the child gathers most of their valuable experiences. Like Dewey, Montessori attempted to bring about changes in the system of education, not only in her own Italy but also to anyone around the world willing to adopt her method.

Montessori subscribed to a more structured learning environment, one in which children begin formal learning at a very early age. In that she differs from Rousseau, but in ascribing more freedom and less powers to the child there was hardly any difference between them. In Kramer (1976) Montessori argue that freedom within carefully placed limits, and not authoritarian discipline, is the principle of education (p.7), and that is significant because the charges leveled at Rousseau by his critics is that he is willing to let children do as they please without any adult oversight or guidance. That overly simplistic approach to Rousseau on the part of his critics is effectively countered by Montessori when she suggests that this freedom ought to have carefully placed limits.

In Montessori (1912) we find a definition of freedom not as an external sign of liberty but as a means of education, and an overwhelming endorsement of Rousseau's views as it relates to the importance of adults respecting the child's freedom. Montessori describes the children in her school as the promise of things to come, and as the future of the specie: "They are the earnest of a humanity grown in the culture of beauty---the infancy of an all-conquering humanity, since they are intelligent and patient observers of their environment, and possess in the form of intellectual liberty the power of spontaneous reasoning" (p.84, 308).

Although the focus of this book is less power and more freedom, it is important to note that this does not preclude the natural powers the child possesses, not the ones given artificially to the child through misguided adults, powers they often remove at will, rendering the child in a lower position than powerless, it renders the child a victim. It is this natural power that is worth protecting. One such form of power, in Montessori's opinion, is the joy the child experiences when exposed to new knowledge. It includes the power to learn from the environment by means of the senses. In Montessori's opinion these are important forms of power (p.301).

In a direct reference to Rousseau and the similarities between them particularly on the subject of freedom and liberty for the child:

> The school must allow freedom for the development of the activity of the child; if scientific education is to come into being…No one would dare to assert that such a principle already exists in teaching or in the school. It is quite true that certain pedagogues like Rousseau set out fantastic principles and vague aspirations of liberty for the child, but the true conception of liberty is, in fact, unknown to the pedagogues (p.7).

The similarities between Montessori and Rousseau on the question of less power and more freedom is clearly established in their writings and perhaps in the writings of most thoughtful thinkers who have argued for educational reform.

The Swiss theorist Jean Piaget rarely addressed the issue of freedom and power as it relates to the child but we do find in Mooney (2000) references to the similarities that exists between Piaget and Rousseau on the subject of more freedom for the child to play and learn as much as possible through that activity as she argues that Piaget stressed the importance of play as an important avenue for learning. Mooney (2002) also added, "It is largely the influence of Piaget, building on Montessori's work, that encourages uninterrupted periods of play in early childhood classrooms. When children are interested and much involved in a subject, they need teachers who respect this absorption with their work" (p. 62, 73).

In an interview with Jean-Claude Bringuier, author of the noted *Conversations with Jean Piaget,* Piaget argued that in the history of science and the formation of man's mind, determinism has played too much of a role, that there have been too few cross-roads and too little by way of freedom. (p.102) But in Muller (2008) we find a statement from Rousseau that appears to run contrary to the concept of freedom expressed earlier by Piaget, "Social transmission failed to explain why an individual may criticize collective beliefs in the name of human rights and truth, thereby contrasting the universal to the collective, i.e., truth to opinion "(p 12).

There are further contradictions emanating from Piaget in Muller (2008), when he tells us that according to Piaget the child first creates pretend play autonomously, through individual rather than social processes and through interaction with the environment rather than with people. This contrasts with the Vygotsky modeled soviet school that considers play to be essential (p.86).

The extended period for play which Piaget recommends, is the equivalent of Rousseau's concept of more freedom and both Rousseau and Piaget see play as an important part of the learning process. Piaget found no need to deal with the concept of power, concerning the child, because all of his time was spent observing them carefully, starting with his very own. This sharp and acute observation of children, along with his habit of engaging them verbally gave Piaget a special understanding of how they think and behave, thus removing from the equation the element of awe or the habit of conferring unrealistic powers to the child that are neither useful or requested.

Lev Vygotsky's work with children was not unlike that of Piaget as he too spent a great deal of time observing them and recording his observations. It is these observations that culminated in the theory he was able to put together, his noted theories on the subject of child development. The initial similarity Vygotsky holds with Rousseau lies in his embrace of children's freedom for endless play convinced that in play much learning takes play. Vygotsky rarely uses the word freedom but his emphasis on play and its importance inevitably invokes freedom since all play requires a level of individual freedom. Proposing the principle that children learn as much or more from the environment and from each other as they do from books or the curriculum, Vygotsky propose that children should have the freedom to create their own learning by choosing from the curriculum and from various classroom activities, the ones that best suit their needs and abilities. No doubt this level of freedom poses tremendous problems for teachers primarily because of class size and the challenge of evaluating the child's intellectual growth. That level of freedom proposed by Vygotsky must be matched with measurable results since when all is said and done, a child in the school system is expected to learn how to read, perform basic math and arithmetic, and conduct some basic reasoning.

We are told in Mooney (2000), that Vygotsky studied and responded to the works Sigmund Freud, Jean Piaget and Maria Montessori during their lifetime, and that he was greatly influenced by them, and that after graduating from Moscow University Vygotsky became a high school teacher, where his interaction with children and his careful observation of their learning peaked Vygotsky interest in the psychological aspect of the learning process. This careful observation of his own students in the learning process, lead Vygotsky to develop his first theory which he referred to as: The Zone of Proximal Development. Vygotsky described the Zone of Proximal Development as the area or distance that exists between the most difficult task a child can perform without the assistance of an adult or more advanced student, and the most difficult task a child can perform with that assistance (Mooney 2000, p. 83, 89).

Although this report is centered on a comparative between Rousseau and these five theorists/thinkers, the parallel that existed between Piaget and Vygotsky demands a closer look, and perhaps some more work by way of comparison.

Mooney (2000) tells us that Piaget was of the opinion that the child's egocentrism prevents it from perceiving the points of view of others thus making play less effective and more of a lone activity among these children. Vygotsky on the other hand, embraced the value of play as a valuable learning tool particularly when performed in a social setting (pp, 90, 91).

In addition to play and its relationship with freedom, knowledge and information also play a significant role in the development of individual freedom either for a child or an adult.

In a review of the interaction between the social world and cognitive development Lloyd and Fernyhough (1999), argue that scholars have long been interested in the relations between social factors and cognitive development. Lloyd and Fernyhough (1999), tell us that in his early work, Piaget [1923/1959] argued that children below the age of 7are unlikely to benefit from social interaction, given the egocentric nature of preoperational thought. That children are like scientists, working alone on the physical, logical, and mathematical material of their world in order to make sense of their reality (pp. 311,321). Lloyd and Fernyhough (1999), also tells us that Vygotsky on the other hand believed that development, a social process from birth onwards, is assisted by others (adults or peers) more competent in the skills and technologies available to the cultures, and that development is fostered by collaboration within the child's zone of proximal development. They insist that Vygotsky's theory is the clearest example of a contextual theory which says that individual development cannot be conceived outside of a social world, and that social world is simultaneously interpersonal, cultural and historical. In other words, from a Vygotsky perspective one cannot consider social interaction between peers and between adults and children without understanding the historically formed social context within which that interaction takes place. Children's cognitive development is thus not the product of simply biological maturation, nor of interaction between them and others in their environment (pp. 311, 329).

The other term Vygotsky introduces is Scaffolding describing it as the assistance a child receives in his efforts to accomplish a particular task. We learn that the inspiration for the name is drawn from the technique of painters and construction workers who create extra flooring in the air in order to access a certain area in the work field. Vygotsky's ideas were considered controversial, particularly since he developed important psychological theory without any previous or formal psychological training.

Both the Zone of Proximal Development and the concept of Scaffolding introduced by Vygotsky represent examples of greater freedom the child can now access in his or her intellectual growth process.

Mooney (2000) closes this section of Vygotsky by telling us that for some teachers the idea that children can help each other learn is very freeing. They think back on the numerous times they've interrupted excellent opportunities for group learning by calling children to circle time where they are forced to sit and listen. Observing Vygotsky recommendations teachers have come to see that children learn not only by doing but also by talking, working with a peer and persisting until the task at hand is accomplished. It is this freedom to do and to learn through other means that Vygotsky strongly recommends (p. 92).

Erik Erikson is the last and youngest of these theorists introduced in this report and compared to Rousseau around the subject of more freedom and less power. Mooney (2000) tells us that he was born in Frankfurt Germany in 1902, that he was an artist and teacher who later became interested in psychology. Mooney also explains that Erikson's meeting and interacting with Anna Freud; the daughter of the great Sigmund Freud; played a role in persuading Erikson to study at the Vienna Psycho-analytic Institute where he specialized in child psycho-analysis (p.37). In a statement that may best define the notion of more freedom and less power Mooney (2000) presents Erikson telling us that parents, teachers and caretakers should learn to accept the child's swing between independence and dependence, and reassuring them that both are okay (47).

To the subject of freedom Erikson dedicates the third of his eight stages of development, pointing to the child of three to six years old, a period in which we encourage their fantasy, their curiosity and imagination. Like Rousseau, Erikson refers to this period as a time for play, not for formal education although in Rousseau's case this period is far more extensive.

The other group to which Erikson dedicates much of his attention on the subject to freedom, are the adolescents, particularly those within that group that may be confronting identity problems. It is at this stage between the ages of twelve and eighteen Erikson claims, that the child is confronted with a variety of social and moral issues. It is here that Erikson (1980) introduces us to the concept of a psychosocial moratorium. In this moratorium, the adolescent will use experimentation to grant themselves a prolongation of the interval between youth and adulthood with the purpose of finding a niche in which he or she may fit comfortably (175). Taking some time out, traveling to Europe, taking some time out to smell the roses or just taking some time out to get to know themselves and formulate their own ideas about the world, with ideas removed from that of their parents, peers or their particular culture group. In the final analysis this might be the strongest advocacy for freedom by any of the five theorists presented in this report.

There is a genuine sense of honesty that emanates from those committed to the development of children, and this palpable honesty has the ability to maintain the interest of scholars and researchers on the subject. It can be said, without any attempt at boldness, that the future of the specie is largely dependent on the proper development of those who constitute future leadership. The survival of the specie for the foreseeable future may not be in question, what may be questionable is the quality of life in this projected future, and this is very much dependent on the attention we pay to the development of children.

The five theorists presented here have done their part, and no doubt future theorists will continue to modify and improve on what those before them have worked so hard to put together.

VIOLENCE TOWARDS CHILDREN

After exploring some of the psychological reasons for parents' violence on children, this chapter will deal with the mechanism that allows some parents to reject the long-standing custom of disciplining their children by way of violence. Intended to be an active, functional document, the research will explore the persuasive language and instruments to be used in creating some ambivalence in those parents and other adults that steadfastly stand by the old customs of submitting children to adult's violence.

The epidemic of violence sweeping the world today has succeeded in tearing apart families and communities, stretching the resources of many social service agencies to its limits. This epidemic has also rendered vast sections of many countries around the world off limits and un-inhabitable to anyone with civilized behavior. Interestingly enough, this is not the first generation of humans grappling with wholesale gratuitous violence. Since the invention of weapons, humans have created extraordinary and sophisticated ways of hurting each other, a behavior that leads to many armed conflicts, and that is also intensified during the course of these conflicts. This wholesale violence took on new dimension with the invention of gunpowder, a product that became synonymous with facilitating killing. Repeated statements that weapons are not the cause of violence may have numbed our senses to these issues of violence and may have also reduced the responsibility we each have for studying this phenomenon, defining its cause and eventually bringing it under control.

The first question that is asked in this report is whether violence is just a by-product of the irresistible desire the strong has for taking advantage of the weak when the opportunity presents itself? Would this then be a confirmation of the violent nature of the species and that violence may be coded in our DNA?

The answers to these questions may never be found, but their absence behooves those of us who care about the future of the species to come together in the search for, or the formulation of.

Sigmund Freud, the celebrated father of psychology and psychoanalysis, dedicated little or none of his time or energy to the study of the cause of violence in humans, despite the fact that a massive war; and the horrible cruelty leading up to another; took place in his lifetime. Even with two world wars behind us, the species failed to produce an industry designed to steadily and consistently analyze the cause of violence in humans. Not until recently did it begin to dawn on a few psychologists that the cause of violence may be deeply rooted in childhood, that this strong desire to feel superior to other human beings may be really the cause of nearly all the misery we've heaped on each other since the dawn of civilization. The idea that these concepts are generated in us in our childhood years should be enough to give pause and take stock at the way we treat those human beings whose care we have been entrusted with.

This book is expected to be a significant part of the ongoing discussion on the subject of violence towards children as it implores the energy and engagement of everyone who shares the dream of a future world free of violence, convinced that the sacred concept of non-violence towards children begins in the home with the child. Non-violence in words, non-violence in deeds, and non-violence in actions.

The most unfortunate thing with the issue of the psychological reasons for violence towards children is that so few resources are available on the subject; it is almost as if writers and researchers have purposely stayed away from this extremely sensitive subject. Yet, it is the argument of so many that without willingness for open dialogue on the subject, the reality of our daily lives as it relates to violence, has reduced chances of ever improving.

Radda Barnen the Swedish version of *Save the Children* issued a pamphlet entitled *Hitting People Is Wrong - and Children Are People Too*. The pamphlet was later published by EPOCH (End Physical Punishment Of Children) an informal alliance of organizations, which share the aim of ending all physical punishment of children by education and legal reform. The pamphlet is a list of six answers describing why parents hit children and in the final of these answers it states: "Many parents are under stress from difficult socio-economic conditions. Forbidding physical punishment would add to that stress and should await better standards of living." The pamphlet goes on to say—"This argument is a tacit admission of an obvious truth: physical punishment is often an outlet for the pent-up feelings of adults rather than an attempt to educate children" (p.1).

The article goes on to argue that in most parts of the world parents urgently need more social and economic support than they get, but they refuse to accept this behavior as justifications for venting their frustrations on children. They assert that children's protection from physical punishment must not be dependent on improvements in the socio-economic arrangements in their parents' lives. What is worse, they argue, hitting children is seldom an effective stress-reliever and cite as evident that most parents who hit out in temper experience guilt and wish that they could find other ways of disciplining their children.

The argument made by the pamphlet is that alternatives to physical punishments are not different punishments but an approach to 'discipline' which is positive rather than punitive, and they cite research showing that effective control of children's behavior does not depend upon punishment for wrong-doing but on clear and consistent limits that prevent it. They explain that adults modeling and an explanation of the behavior they would prefer for the child seems to have a more positive effect on curbing the child's behavior (p.2).

A more radical approach to this subject was offered by Jordan Riak head of the Parents and Teachers Against Violence. In the 1992 issue of the organization's newsletter, Riak published an essay entitled Plain Talk About Spanking and in the essay Riak argues that many spankers are habituated to the practice because it provides them with an instant outlet for their feelings of frustration and anger - not because they've found it an effective way to improve a child's behavior. The danger of this he argues, is that violence, by its very nature, tend to escalate as it is indulged in, thus making it impossible for there to be a safe way to hit a child (p.1).

Riak (1992), then goes on to make the connection between spanking and sexual molestation telling us that spanked children learn that their bodies are not their personal property, and that allowing someone else to do as they please with their bodies opens the gate for that someone or others to do the same or even more, and that even their sexual areas are subject to the will of adults. The child who submits to a spanking on Monday is not likely to say no to a molester on Tuesday. So no matter what else violent parents think they are accomplishing with their behavior, they are setting children up to be easy prey for predators (p.2).

The other area in which those parents try to justify their behavior is arguing that the buttocks is safe because of its meaty structure but Riak then commented that medical science has long recognized and documented in great detail how being struck on the buttocks can stimulate sexual feelings. Riak (1992), makes it clear that located deep in the buttocks is the sciatic nerve, the largest nerve in the body and that a severe blow to the buttocks, particularly with a blunt instrument, could cause bleeding in the muscles that surround that nerve, possibly injuring it and causing impairment to the involved leg. Riak (1992), adds that a blow to the buttocks can cause injury to the tailbone (coccyx) or sacrum. It sends force waves upward through the spinal column possibly causing disc compression or compression fractures of vertebral bones. And as far as the old claim that God or nature intended that part of the anatomy for spanking Riak argues that that claim is brazenly perverse since no part of the human body was made to be mistreated (p.4).

The tragic consequence for many children who have been punished by spanking, according to Riak (1992), is that they form a connection between pain, humiliation and sexual arousal that endures for the rest of their lives. Riak then proceeds to introduce David Bakan, author of *Slaughter of the Innocents*, in which Bakan wrote:

> The buttocks are the locus for the induction of pain in a child. We are familiar with the argument that it is a safe

84

'locus' for spanking. However, the anal region is also the major erotic region at precisely the time the child is likely to be beaten there. Thus, it is aptly chosen to achieve the result of deranged sexuality in adulthood. (Bakan 1971 p. 113).

Riak (1992), continues to present the argument for no-spanking telling us that the pornography and prostitution industries do a thriving business catering to the needs of countless unfortunate individuals whose sexual development has been derailed by childhood spankings. If we put all other considerations aside, this should be reason enough never to spank a child (p.5).

The excuse that so many schools give that the hands are safe for hitting is also refuted by Riak (1992), stating that his research has revealed that the child's hand is particularly vulnerable because its ligaments, nerves, tendons and blood vessels are close to the skin, which has no underlying protective tissue. Striking the hands of younger children is especially dangerous to the growth plates in the bones, which, if damaged, can cause deformity or impaired function. Striking a child's hand can also cause fractures, dislocations and can lead to premature osteoarthritis, he argues. Many of us have also become familiar with the shaking baby syndrome, but not everyone has. Shaking a baby that is crying annoyingly seems innocuous to many uninformed parents specially those parents in the lower strata of our society, and those in the developing world, so the damage or death to those children will forever go undetected (p.5).

Riak (1992), ends his comments by telling us that we should not be surprised that many youngsters reject the adult world to the degree they believe it has rejected them. Nor should we be surprised that those who throughout childhood have been recipients of violence will become dispensers of it as soon as they are able. Some teachers work tirelessly to curb violence-impacted children's aggressiveness, to instill trust which those children lack, and to redirect their energies in positive directions but that is a daunting task even for the most dedicated and best prepared teachers since it requires extraordinary resources currently inaccessible to the current public-school systems. School dropout, addiction and delinquency would cease to be major problems if only it were possible to persuade parents and other caretakers to stop socializing children in ways likely to make them antisocial and/or self-destructive (p. 6).

RADICAL NON-VIOLENCE
A NEW PARADIGM

In our times the term radical has taken on new and more perverse meaning for even though its *Latin* etymology is Radix or root, today the term has come to describe the most dogmatic, violent and narrow-minded elements within a particular religion. Despite that unfortunate reality, we are not yet willing to give up on the term, because it allows us to understand the origins and root causes of any issue.

Radical non-violence begins with an extraordinary respect for all living things, especially the small and defenseless. We carefully avoid the term love because that too has been misunderstood and misused. You simply cannot love someone and mistreat them at the same time, another word must be invented if we are to mix mistreatment with love. Respect however, carries no emotions, focusing solely on the intrinsic value of the person or living being regardless of its size. It is virtually impossible to hurt; physically or otherwise; that which we happen to respect.

Real non-violence represents a paradigm shift in the ways that human beings relate to each other. In this case it may also be referred to as radical non-violence because of its focus on the root causes of the anger and vitriol that seems to characterize human interactions.

Humanity has had three renowned non-violent advocates: the Palestinian Jesus Christ, the Hindu Mahatma Gandhi and African American Martin Luther King, and it is likely that there be a great many others dedicated to the cause of non-violence who we simply did not become aware of. Violence towards children was as prevalent in their times as it was in any other, yet in none of these cases have we read of a denunciation of that violence that is perpetrated against children or a recognition of the damaging effects of this violence. In his 1974 publication *The History of Childhood* Lloyd deMause reminds us that at the time of Jesus' birth, infanticide was a common practice throughout the world, including the Middle East, and that his brief presence on this planet had no effect whatsoever on this horrifying practice. deMause (1974) reminds us that sealing children in the walls and foundations of bridges and buildings to strengthen the structure was also common. From the building of the wall of Jericho to as late as 1843 in Germany (p.27). Likewise, we find none of Gandhi's writings reflects a specific interest in the safety and protection of children, the same is true for the writings issued by and about Martin Luther King.

This new approach to non-violence is unprecedented, it gives personhood to the child and it calls on civilized people to address the child's behavior clinically utilizing the techniques developed after more than two centuries of psychological practices.

It is the responsibility of every legitimate government to guarantee the safety and security of all of its citizens, particularly the most vulnerable. Fortunately, state and local governments around the world have been responding to the United Nations 1978 Declaration of the Right of the Child which says in essence that all children deserve to live in safety and security.

The goal is non-violence in the home, particularly with respect to children, and we believe that governments have it within their powers to provide incentives for parents who sign on to this agreement. Tax breaks, food surplus programs, high school diplomas or college credits for those who successfully comply are but a few of the options that with some creativity, government can make available to these parents and caretakers who have done their part in helping to create better citizens. However, this will not happen by itself, with this priority in mind, future organizers will learn to bring pressure to bear on both elected and appointed officials at all level of government to endorse these creative ideas whose ultimate effect will drastically reduce social problems and create a more functional citizenry.

Partnering with parents is essential to providing greater security for children, but for those parents who are already borderline dysfunctional or those whose method of discipline is deeply engrained, it will require a great deal of persuasion to convince them of the benefits of a new approach to raising their children. Children who have been raised successfully without corporal punishment may be the ambassadors for engaging other parents in an effort to persuade them as to the benefits of this non-violent approach to child rearing, for if parents become convinced that their children can do as well without the use of violence there is a chance they will consider this method particularly when the by-product of this new method is in their presence. Many of the parents who need to be dissuaded from the old methods are themselves functionally illiterate, so volumes of written materials may not be useful in this case, it will require a core of committed organizers that includes adolescents, concerned mothers, seniors and anyone else with the ability to persuade, to reach out to these parents and caretakers and convince them as to the validity of this new method.

The concluding thought in this chapter is that only a, well-structured campaign against violence can really begin to force parents and other caregivers to take a second look at their behavior, and in the process, to contemplate other avenues for raising their children. It is expected that at some point the United States will join the other twenty-nine nations that have passed laws against corporal punishment, and when it does, this will become an added component to the persuasive mechanisms currently in place to guard the safety of children and the healthy survival of our society. For now, we can view this as the most compelling and most satisfying of all endeavors.

FEMALE GENITAL MUTILATION

With the exception of outright murder, which was committed systematically against children just a few short centuries ago, Female Genital Mutilation is regarded by its victims as the worst form of violence.

It could be said that outside of crimes like rape, incest and brutal beatings in which the child survives physically, hardly anything compares with this supposedly well meaning but grossly misguided adults who butcher the private parts of an innocent little girl in the name of culture. In fact that is the excuse commonly offered for the lack of intervention in this practice, making an assault on a person acceptable only by invoking the word culture. Described by the literature and by the victims, only rape and murder can surpass the horrors of Female Genital Mutilation. Frye (2004) commented although Female Genital Mutilation is sometimes referred to as female circumcision, perhaps as a way of sanitizing the concept and to make it more acceptable in the minds of the readers for the horrors it entails would be an assault on the senses. She chose to use the full expression Female Genital Mutilation because it demands of the reader and the researchers a complete understanding of the horrors some preteen girl is undergoing even at the moment they are consuming the information. Female Genital Mutilation is primarily practiced in parts of Africa, the Middle East, Asia and some islands in the Pacific some of it is now being seen in the United States, the United Kingdom and other Western countries as immigrants bring with them their degrading practices. She offers the World Health Organization's 2008 report that approximately 140 million women and girls worldwide have undergone Female Genital Mutilation since they began keeping track, and that even though its prevalence in the United Stated is unknown, given the 2004 reports from the Center for Reproductive Rights, women in Western countries are currently at great risk of being subjected to these practices. What is interesting is that the World Health Organization has concluded in its 2008 report that there are no health benefits to this practice and that it constitutes in every sense a violation of human rights. While some say that Female Genital Mutilation should be prohibited globally because of its barbarity and because it constitutes one of the worse forms of child abuse, there are those who argue that it is a rite of passage that needs to be observed in order to preserve cultural identity. (Frye 2004, Barstow 1999).

Frye (2004) presents the history of Female Genital Mutilation as a practice that dates back more than 5,000years, one that may have had its origins in ancient Egypt. Details of how it came about were not offered but he stated that from the examinations of ancient Egyptian mummies, there is evidence that Female Genital Mutilation was practiced routinely.

Religious people supporting FGM claim that their religious texts require these physical alterations as they are a necessity in securing a woman her standing within her community. FGM goes beyond circumcision in removing the male aspect from a woman and further instills her femininity as she defines herself as a woman that is now more enhanced, docile and obedient as a result. Currently, African culture works to uphold a male-dominated society. This precedent is not in line with the goals of the lAC as its goals work to establish a more equal role for both males and females. Not wanting to alienate itself from the general public, the lAC uses mildly aggressive tactics to induce cooperation and advance the woman's role in African culture (Frye 2004).

Despite the perception of Female Genital Mutilation as a foreign practice, its other name, Clitoridectomies, was practiced routinely in the United States and Europe to treat lesbianism, hysteria, melancholy, epilepsy and excessive masturbation as recently as the late nineteenth century, in fact she added, the practice continued in the U.S. until the late 1930s when an outrage public demanded its abolishment. The author added that despite the belief by many that Female Genital Mutilation is a religiously related practice, in actuality the practice is not associated with any specific religious faith and there are no religious scripts that proscribe this barbarous act. It has been illegal in Britain since 1985, she tells us but it is still practiced secretly in some immigrant communities, and when that proves too risky, girls are sent abroad to have the procedure thereby skirting the legal restrictions. In 1997 and again in 2008, the World Health Organization issued a joint statement with the United Nations to support increased advocacy for the abandonment of the practice of Female Genital Mutilation enlightening the public as to the intricacies of the practice. They describe it as characterize by four distinct types: The first is a Clitoridectomy, a procedure that partially or totally removes the clitoris. The second type, Excision, is the partial or total removal of the clitoris and labia minor which may include incision of the labia major. The third is, Infibulations, the narrowing of the vaginal opening through the creation

93

of a "covering seal," which is formed by cutting and repositioning the inner, and sometimes outer labia. Sometimes this includes removal of the clitoris and sometimes it does not. A fourth category is preserved for other, which includes any harmful, non-medical procedure done involuntarily to the female genitalia such. (Frye 2004).reports that in the past Female Genital Mutilation has been carried out by a female circumciser, that the procedure was done without anesthetic, antiseptics, or antibiotics often in unsanitary conditions utilizing instruments such knifes, razors, pieces of glass or sharpened stones, instruments that were likely to be dirty, having been used on other girls and never sanitized. She goes on to report that even though the procedure is being done by trained medical personnel in a few places around the world, it continues to be an act that results in permanent physical mutilation to the child, one that changes body functions forever, with immediate complications and extreme pain in all cases, not to mention shock, hemorrhaging, tetanus, sepsis, urine retention, open sores and injury to nearby genital tissue. Other long-term consequences include recurrent bladder and urinary tract infections, cysts, infertility, need for later surgeries, increased risk of childbirth complications and newborn deaths. In nearly all cases, a host of surgeries may be required in later years to allow for sexual intercourse or for a pregnant woman to deliver a child. On a nonphysical level the psychological consequences of this traumatic experience is one that is yet to be fully analyzed, but reports of post-traumatic stress disorder, anxiety, depression and psychosexual problems are fairly common. (Frye 2004) ponders the psychological dilemma that allows these victims to subject other little girls to the same procedure just a few short years after their own trauma. Arguments such as female rite of passage, custom and tradition, the clitoris is dirty or evil, the clitoris will grow too long, non-excised women will be barren, the clitoris causes male impotence, it insures virginity and chastity, women who have not undergone the procedure are rejected as marriage partners, it increases a woman's femininity, it prevents social ostracism, it controls the female sex drive, it prevents lesbianism, it ensures paternity, it calms the female personality, it defines cultural identity, it encourages cultural cohesion; have long been offered to justify the procedure. Those who argue for the abolition of this barbarous procedure are accused by its defenders of ethnocentrism, cultural imperialism and cultural imposition, as they assert their rights to continue subjecting defenseless little girls to this ancient procedure. Frye (2004) alludes to the procedure as a gender-specific human rights violation and a form of child abuse reflecting a deep-rooted inequality between the sexes, and an extreme form of discrimination against women, adding that the fact that the practice is specific to certain cultural groups should not excuse it from international scrutiny. Healthcare professionals worldwide are charged with the responsibility of taking a stand on this issue, and to understand the abhorrence of this practice. The author concludes that culture should be progressive beneficial and not static or retrogressive to the detriment of those who by mere accident of birth belong to a practicing clan, that certain cultural rituals found at odds with the fundamental rights of individuals are being discarded as the global village shrinks in the face of rapid technological advancement. She argues that Female Genital Mutilation should incur global abolition the same way other anachronistic global tradu cultural practices have, *ie*; that it must be abolished because it is a violation of human rights. In an apparent invocation of cowardice Barstow (1999) charged that the human species has habitually exhibited cruelty towards its weaker members, notably women, children and the elderly (p. 73) and goes on to say that development in human rights around the globe has brought much attention to the wanton violence, sexual violence, deprivation, oppression and outright murder committed against women everywhere (p.155). With the adopting of the Convention on the Rights of the Child in 1959 the general assembly of the United Nations has cast its lot on the side of children irrespective of cultural norms or beliefs. The measure states explicitly that all nations must take measures to abolish traditional practices that are damaging to the health of children.   In a letter to the editor in the March April (1997) issue of the Canadian Journal of Public Health, men pay high dowry for women who have undergone Female Genital Mutilation as they've acquired a woman who had not been penetrated by other men before him, but it is obvious they're not   aware of the health consequences, and why their partners suffer from multiple health problems.

EMPOWERING CHILDREN

The concept of empowerment is a relatively new one in our lexicon as traditionally, people never felt they had the right to be empowered. This concept was introduced in a massive scale by the young men and women leading the sixties movement as they demanded power for the people. A people that until that point had been completely powerless, and in nearly every sense continues to be just as powerless as they were a half a century ago. The question then continues to be: How do we go about empowering the powerless? A few thousand years ago some shrewd operators took on the mantle of Kings and Queens, and they somehow convinced the masses of the time, of their intrinsic superiority. This hypnotic control over the masses by an insignificant few lasted for thousands of years until the French revolution and others of its kind initiated a process in which that power would be arrested from those who claimed it for themselves. The process was bloody and untidy but eventually the masses came out on top. Some years later, we had some semblance of what was supposed to be a socialist revolution, one that was really designed to restore power to the masses. However, personality worship had already replaced the worshipping of the unseen and the unknown, giving individuals the opportunity to pose as gods, exercising the powers and privileges previously preserved for gods. That reality has lasted through our times albeit in different versions.

The revolt of any oppressed group grants the rest of the population an opportunity to crawl out from under their own oppression as someone else is taking their lives into their own hands by confronting power. What is real however, is that these revolutions bring with them some confusion and some lack of clarity that makes it possible for them to be co-opted and rerouted before their potential or intended objectives can be accomplished. There is a growing atheist movement and some feel this may actually be the long-awaited catalyst, but this group lacks the force and the passion to bring about any meaningful change. In addition, this group has not had to overcome oppression and discrimination brought on by their own natural identity. This book makes the claim that only an open challenge to the childish notions that drives all belief systems can really empower the masses freeing oppressed groups from the tyranny of the majority.

RELIGIOUS INDOCTRINATION
AND ITS IMPACT ON THE CHILD

Before embarking on the subject that encompasses the form of indoctrination most of us are familiar with; that is religious indoctrination; it is important to attempt a definition of the term itself and the various ways in which it has been utilized in the past. There is no question that over the years we have had a variety of definitions for the term. Barrow et, al (2010) comments that much of it depends on the academic level of the person offering the definition, or their political or intellectual inclinations. In offering a generalized and perhaps more generic definition for the term, we can safely say that it is a process in which coercion, persuasion and often compulsion, are used to introduce an idea or belief system to an individual or group of individuals. In this definition, a key component to the process invariably, is the absence of critical examination or questioning on the part of the new adherents. Barrow et al (2010), comments that the term has obtained a strong pejorative tinge in the educational discourse particularly because of the absence of critical thinking and questioning the term is associated with. Barrow et al (2010) tell us that the term is often contrasted with the grand ideas of autonomy and open-mindedness that our society has come to expect from any form of positive instruction (p. 279).

Religious indoctrination on the other hand, involves a particular effort to have its adherents submit to a prescribe set of narrow and irrational beliefs that invariably involves a concept of god as an image or person to be adored, to be in awe of, or to be

fearful of. The proven and tested method is to implant these dogmas in the minds of the individual long before he or she achieves
the capacity to discern the process for themselves, thus preempt-
ing all possibilities for resistance.

In their own analysis of this subject, Barrow and Woods
(2006) introduce us to the term "unshakable belief", related to the
intentions of the indoctrinator's intent. Getting someone to believe in a proposition that is lacking in proof or evidence is at the
heart of the process of indoctrination. Barrow and Woods (2006)
remind us that the essence of indoctrination is not to be found in
the degree of conviction but in the blind unshakable commitment.
It possesses something that is beyond argument, and beyond reasoning, that is its position as being antithetical to education (p.
71). Barrow and Woods (2006) tell us that there are things in the
world that are different from education and even incompatible
with it in certain areas, but indoctrination is education inverted.
It involves denial of the value of rationality, a real phenomenon
that we must contend with for reasons that are too numerous to
list (p. 71). To be honorably committed to a cause is one thing,
but indoctrination is clearly another level of functioning, one that
requires much more of our attention. The term unshakable belief
reminds us of the extraordinary efforts the neurological system
had to make to accept irrational thoughts.

In contemplating the Christian symbol of a gentle Jesus, meek
and mild one has to wonder if the violent scenarios described in
Reuters (2004) could be tolerated within Christianity. The stories
of violence described in the first half of the Christian bible known

as the Old Testament are a possible example of the similarity between the Abrahamic faiths and why this phenomenon may not be inconceivable and violence

One is compelled to wonder about the effects strong and violent biblical statements have on the minds of children that are under the indoctrination of religion, particularly the Abrahamic faiths.

Children brought up in the Christian tradition and instructed to read the bible are compelled to make sense of this aspect of their religion. The notion of an all-loving god comes into question for them and in many cases the entire foundation of their belief system crumbles.

The writings in these chapters and verses vary according to the version of the bible the person may be in possession of, thus, one is free to consult the versions of their choice or a few versions, for that matter, for the purpose of comparison. These writings contained in the bible have been condemned by some for their irrationality and dismissed by others as insignificant but rarely are they analyzed for the effects they have on the minds of the masses, particularly on the children's.

RELIGION IN HUMAN RELATION

The task of exploring the role of religion in human progress or non-progress is a rather difficult one, one that requires an overwhelming degree of what is referred to as intellectual honesty, that is, a willingness to explore areas that are considered off limits to most humans, and confront what are considered established norms. Because of that, realistic research on a subject such as religion is all the more challenging and for it we must thank the few daring researchers, who risk so much to stare down and confront established norms. Old friends for daring to challenge the sacred have cast many off as pariahs and others were shunned by colleagues and other professionals, as they embarked on a subject matter that made them uneasy. To add to that, the few who would in the past dare to venture into these delicate areas would often do so denouncing the established order as well as its administrators, by-passing the critical mass of thinkers required to have some impact on the subject. This denunciation and open confrontation has led to confusion rather than clarity, particularly for those tapped under the influence of a rather powerful socio-political infrastructure. The result of this of course, is limited research material on a subject of this nature. There are researchers and academicians dedicated to this subject but their numbers are limited, and since the first label attached to anyone who dares to take on this subject is that of atheist, there is a considerable reduction in the pool of intellectuals willing to go that far.

No other generation in the past has had this much exposure to scientific facts and evidence surrounding the presence of humans in this universe. If for no other reason than that, we owe it to

102

ourselves to persuade families of all religious inclination or none at all, to guarantee that these facts are not hidden from their children.

PSYCHOTHERAPY
AND THE HEALING PROCESS

This book attempts to present a comprehensive review of the process we have come to know as psychotherapy, its history and the contribution it has made to overall human happiness and the advancement of the species. Dr. Jerome Frank's 1963 publication *Persuasion and Healing,* in which he compares the various therapeutic approaches of the past fifty years, will serve as a point of departure as I explore the multicultural dimension of the psychotherapeutic process, and the effectiveness of its application in a cross-cultural setting.

Reasoning and introspection therefore, it can be argued, are the foundations for philosophy, as humans began to ask questions regarding themselves and their environment, their origin, and their destiny. This deep thinking also gave way to the discipline we have come to know as psychology, a Greek term for the study of the soul. (*Psyche = the Soul Logos = Study*).

A clear understanding of what we have come to know as the helping profession is important if we are to make these services available to a larger number of people in our society, recognize the generalized emotional malady that touches nearly everyone and eventually improve the efficacy of the healing that so many are so desperately in search of.

The blurring of the line that separates facts from fiction creates a never ending and violent conflict inside the mind of many individuals, and this conflict leads to the type of confusion that effectively serves the interest of those with answers for which there are no evidences. These individuals, or groups as the case may be, use this confusion to their advantage promoting their brand of truth thus increasing the sense of guilt and confusion among the confused and most self-harming behaviors are the result of confusion.

Psychologists and philosophers are yet to define ignorance and manipulation as the nemesis threatening sanity, security, world peace and individual happiness. Until they do, the revolving doors of therapeutic and behavioral centers everywhere may be swinging perpetually.

Distinguishing the two and acting both forcefully and legally in the construct of that distinction is the psychologists' greatest challenge.

But for a thorough understanding of the practice we have come to know as psychotherapy it is important to take a closer look at what we have come to know so far as human history, bearing in mind all of its contradictions and polemics surrounding the topic.

Homo sapiens is the term used to describe the creatures invested with the ability to reason, with the capacity for introspection and with the ability to resolve problems. To that, we may add language, although a voice box and vocal chords came some time after Homo sapiens developed basic problem-solving skills.

With the rapid changes in technology and a transient society moving at speeds that very few are able to keep up with, there may never have been as great a demand for the therapeutic process and the sense of wellness it is expected to provide. Everyone from the high-powered executive to the housewife, the lawyer, the builder, the doctor, and professions of all sorts, are all in need of the assertiveness and the coping mechanisms provided by those who are engaged in the helping profession.

Psychology is still in the process of defining itself and fending off critics who still attempt to write the profession off as quackery or pseudoscience. However, for those who have assisted another human being in becoming even slightly more functional in a world overrun by madness, uncertainty and confusion, their feelings about this profession is an entirely different one.

These brave soldiers in the battle for clarity, understanding and self-realization for others, are the ultimate helpers, in need of no gratitude or rewards as they commit to their best efforts in improving the lives of those who seek them out for assistance.

ALICE MILLER (1923-2010)

We learn from Miller (1984) that this giant of the field of psychology was born to a Jewish family in Poland and was trained as a psychoanalysis in Switzerland obtaining her doctorate in 1953.

In 1980, after having worked as a psychoanalyst and an analyst trainer for 20 years, Miller suspended her practice and dedicated herself to exploring the issue of childhood. She was critical of both Sigmund Freud and Carl Jung for ignoring this issue and several others in their practice and in their writings. Her first three books originated from research she took upon herself as a response to what she felt were major blind spots in her field. However, by the time her fourth book was published, she no longer believed that psychoanalysis was viable in any respect. In 1985 Miller wrote about the research from her time as a psychoanalyst: "For twenty years I observed people denying their childhood traumas, idealizing their parents and resisting the truth about their childhood by any means."

According to Miller (2002), despite all of Stalin's power, he spent his lifetime in fear of his father. Regarding Hitler, she stated he believed that the annihilation of millions of people would free him of the tormenting fear of his violent alcoholic father. *Poisonous Pedagogy,* in her opinion, contributed a great deal in millions of children and adults supporting Hitler's atrocities without experiencing a sense of horror. Hitler and his cohorts, she argued, were part of a generation of children who had been exposed to brutal physical correction and humiliation, and who later vented their pent-up feeling of anger and helpless rage on innocent victims. Safe in the knowledge that they were doing so with the Fuhrer's blessings, they were finally able to give free reign to those feelings without fear of punishment. Her claim is that wherever cruelty and humiliation are a part of parenting, those methods will be reflected in the behavior of young people.

Miller (2010), further states that in the name of good parenting millions of children all over the world are subjected to some of the worst form of violence. Mao Tse Tung was the son a strict teacher who set out to drum obedience and wisdom into him with the aid of severe physical correction. Mao later attempted to install this same philosophy in his country at the cost of more than thirty five million lives. Like *Hitler, Joseph Stalin* was exposed to incredible brutality as a child. Stalin's brutal treatment at the hands of an alcoholic father made his childhood pain with dread of being killed during one of his father's outbursts. As an adult, he had the power to fend off that fear by humiliating others.

In short what we learn from Miller (2010) and nearly all of her publications and interviews is that worldwide violence has its roots in the fact that children are beaten all over the world, especially during their first years of life, when their brains become structured. She said that the damage caused by this practice is devastating, but this subject is hardly discussed in our society. She adds that the suppression of their natural reactions like rage and fear generated by their inability to defend themselves against the violence inflicted on them forces them to discharge these strong emotions later as adults against their own children or whole nations. The example she often points to is Adolf Hitler whose childhood was filled with violence coming from an alcoholic father. Violence towards the child in the form of beating and humiliating not only produces unhappy and confused children, but also confused, irrational adults who in turn create dysfunctional and irrational societies. Miller concludes that only when our societies become aware of these complex dynamics will we break this insidious chain of violence.

The concept of Moral Injury may be worth looking into, particularly since it has everything to do with notorious acts of violence that are perpetrated against large numbers of human beings. Exploring whether some sense of guilt or remorse is experienced by the perpetrator is the psychological phenomenon that this term was originally intended to deal with.

Vargas (2013) defined Moral Injury as that complex feeling that arises in the individual after perpetrating, failing to prevent, or bearing witness to acts that is beyond that individuals deeply held moral beliefs and expectations. The literature on two of the world's worst mass murderers reveal that they were deeply religious, or at least were brought up in religious households. The question that is asked is whether the element of Moral Injury played a role at any point in these men's life since according to the literature Moral injury can lead to serious distress, depression and suicidality. In April 1987 Miller announced in an interview with the German magazine *Psychologie Heute* (Psychology Today) her rejection of psychoanalysis. The following year she cancelled her memberships in both the Swiss Psychoanalytic Society and the International Psychoanalytic Association, because she felt that psychoanalytic theory and practice made it impossible for former victims of child abuse to recognize the violations inflicted on them and to resolve the consequences of the abuse,[10] as they "remained in the old tradition of blaming the child and protecting the parents.

Miller blamed psychologically abusive parents for the majority of neurosis and psychosis. She maintained that all instances of mental illness, addiction, crime and cultism were ultimately caused by suppressed rage and pain as a result of subconscious childhood trauma that was not resolved emotionally, assisted by a helper, which she came to term an "enlightened witness." In all cultures, "sparing the parents is our supreme law," wrote Miller. Even psychiatrists, psychoanalysts and clinical psychologists were unconsciously afraid to blame parents for the mental disorders of their clients, she contended. According to Miller, mental health professionals were also creatures of the poisonous pedagogy internalized in their own childhood. This explained why the command "Honor thy parents" was one of the main targets in Miller's school of psychology.

Miller called electroconvulsive therapy "a campaign against the act of remembering." In her book, *Abbruch der Schweigemauer* (The Demolition of Silence), she criticized psychotherapists' advice to clients to forgive their abusive parents, arguing that this could only hinder recovery through remembering and feeling childhood pain. It was her contention that the majority of therapists fear this truth and that they work under the influence of interpretations culled from both Western and Oriental religions, which preach forgiveness by the once-mistreated child. She believed that forgiveness did not resolve hatred, but covered it in a dangerous way in the grown adult: displacement on scapegoats, as she discussed in her psycho-biographies of Adolf Hitler and Jürgen Bartsch, both of whom she described as having suffered severe parental abuse. A common denominator in Miller's writings is her explanation of why human beings prefer not to know about their own victimization during childhood: to avoid unbearable pain. She believed that the unconscious command of the individual, not to be aware of how he or she was treated in childhood, led to displacement: the irresistible drive to repeat abusive parenting in the next generation of children or direct unconsciously the unresolved trauma against others (war, terrorism, delinquency) or against him or herself (eating disorders, drug addiction, depression).

Miller (2010) states that worldwide violence has its roots in the fact that children are beaten all over the world, especially during their first years of life, when their brains become structured. She said that the damage caused by this practice is devastating, but unfortunately hardly noticed by society. She argued that as children are forbidden to defend themselves against the violence inflicted on them, they must suppress the natural reactions like rage and fear, and they discharge these strong emotions later as adults against their own children or whole peoples: "child abuse like beating and humiliating not only produces unhappy and confused children, not only destructive teenagers and abusive parents, but thus also a confused, irrationally functioning society. Miller stated that only through becoming aware of this dynamic can we break the chain of violence.

For an example of the calamity we as a species expose ourselves to when we ill-treat children and deny them the healing they deserve we need to go no further than Adolf Hitler. Waite (1971) presented a highly researched and carefully thought-out paper on Hitler. In it he explains that six million Jews were sacrificed to Hitler's personal sense of unworthiness and hyper vulnerability of the body to filth and decay. So great were Hitler's anxieties about these things, so crippled was he psychically, that he seems to have had to develop a unique perversion to deal with them, to triumph over them. He added that: "Hitler gained sexual

satisfaction by having a young woman   squat over him to urinate
or defecate on his head." This was his "private religion": his personal transcendence of his anxiety, the hyper experience and res-
olution of it. This was a personal trip that he laid not only on the
Jews and the German nation but directly on his mistresses. It is
highly significant that each of them committed suicide or tried to
do so. and more than a simple coincidence. It might very possibly
be that they could not stand the burden of his perversion; the
whole of it was on them, it was theirs to live with — not in itself,
as a simple and disgusting physical act, but in its shattering ab-
surdity and massive incongruity with the role he was playing at
the time. The man who is the object of all social worship, the
hope of Germany and the world, the victor over evil and filth, is
the same one who will in an hour plead with you in private to "be
nice" to him with the fullness of your excretions. I would say that
this discordance between private and public esthetics is possibly
too much to bear, unless one can get some kind of commanding
height or vantage point from which to mock it or otherwise dis-
miss it, say, as a prostitute would by considering her client a simple pervert, an inferior form of life. (p. 234). Lloyd deMause, a
collaborator and close friend of Alice Miller has arrived at similar
conclusions. He stated in deMause (1974):

> The history of childhood is a nightmare from which con-
> temporary society has only recently began to awaken. The
> further back in history one goes, the lower the level of
> child care, and the more likely children are to be killed,
> abandoned, beaten, terrorized, and sexually abused. (p. 1)

deMause's continuous reference to psychogenics reminds
readers of the possible relationship that may exist between current socio-economic problems and the treatment meted out to
children for nearly all human history as each traumatized generation of children acts out their repressed resentment and anger as

112

adults. He argues that the central force for change in society is neither technology nor economics, but the psychogenic changes in personality occurring because of successive generations of parent child interactions. Therefore, examine how children were valued and conceptualized over time can help to better understand the emergence and evolution of violence directed at children.

OTTO RANK (1884-1939)

Born in Vienna Austria in 1884, Rank: was 28 years Freud's junior, and what is interesting in this relationship is that Rank looked up to Freud as his adopted father. He was Freud closest disciple and colleague from 1906 through 1926, the formative years of psychoanalytic movement. Freud valued his expertise in art, music, literature, anthropology, history, science, and philosophy. Freud advised him not to go to medical school, but to complete his academic education. Rank obliged and obtained his PhD at age 28 at the University of Vienna in 1912.Freud, only 5.7" tall would affectionately refer to Rank, 5.3" as "Little Rank" in his letters to another of his closest colleagues, Carl Jung who stood above them both at 6 feet tall. In 1935 after a lifetime as Freud's protégé and follower, Rank left Europe and took up residence in the United States. He was already becoming disenchanted with the movement, criticizing it at being stagnated, and excessive in its attempts to psycho-analyze everything almost to an extreme. Becker (1972) offers us one of Otto Rank's most famous quotation uttered as he was beginning the process of breaking away from Freud and psychoanalysis:

> Suddenly....while I was resting in bed it occurred to me what really was (or is) Beyond Psychology. You know what? Stupidity! All that complicated and elaborate ex planation of human behavior is nothing but an attempt to give a meaning to one of the most powerful motives of behavior namely, Stupidity.! I began to think that is even more powerful than badness, meanness—because

many actions or reactions that appear mean are simply stupid and even calling them bad is a justification (p 251).

The relationship between Freud and Rank lasted 20 years and the two men; it is fair to say; became a lot better as a result of it, despite the difference in age. It is also interesting to know that they both died in 1939. Freud on September 23rd and Rank on October 31st of that year.

ERIC FROMM (1906-1980)

Eric Fromm was perhaps the greatest and most faithful of Freud's followers. A prolific writer himself, Fromm dedicated much of his life and energy giving the world a greater understanding of Freud, his times, his personality, and his work. Nowhere is this done more effectively than in his book, *Sigmund Freud's Mission*. Here Fromm gives us a personal look at Freud and all that is connected to him. He was Freud's junior by 44 years, but Fromm did more detailed and honest study about Freud than any single individual. As his biography suggests, Fromm's theory is a rather unique blend of Freud and Marx. Freud, of course, emphasizes the unconscious, biological drives, repression, and so on. In other words, Freud postulated that our characters are determined by biology Marx on the other hand saw people as determined by their society and most especially by their economic systems. He added to this mix of two deterministic systems, something quite foreign to them: The idea of freedom. He allows people to transcend the determinisms that Freud and Marx attribute to them. In fact, Fromm makes freedom the central characteristic of human nature. There are; Fromm points out; examples where determinism alone operates. A good example of nearly pure biological determinism a la Freud is animals (at least simple ones). Animals don't worry about freedom their instincts take care of everything. Woodchucks, for example don't need career counseling to decide what they're going to be when they grow up; they are going to be woodchucks! Two events shaped Fromm's life, and these two took place long before he met Freud. The first at age 12 had to do with a female friend of the family around 25 years of age. As he described it: She was beautiful, attractive and in addition a painter, the first painter I ever knew he stated. I remember having heard that she had been engaged, but after some time had broken the engagement; I remember that she was almost invariably in the company of her widowed father. As I remember him, he was an old uninteresting and rather unattractive man or so I thought (perhaps my judgment was somewhat biased by jealousy). Then one day I heard the shocking news; her father had died, and immediately afterwards, she had killed herself and left a will that stipulated that she wanted to be buried with her father. This news hit the 12-year-old Eric very hard and he found himself asking many questions, some of which he found answers to later on in his life in Freud. The second event took place two years after with World War I. At the tender age of 14 he saw the extremes that Nationalism could go to. All around him he heard the message: We Germans (or more precisely Christian Germans) are great; they, the English and their allies are cheap mercenaries. The hatred, the war hysteria, frightened him. So again, he wanted to understand something irrational the irrationality of mass behavior, and he found some answers, this time in the writings of Karl Marx. At age 22 Fromm had already received his Ph.D. in Psychology, but the uniqueness of his contribution to psychoanalysis was his own emphasis on Love and Rationale. He argued that the highest value for any human being consisted in finding unity with the world through full development of specifically human capacities of

Love and Reason. All the intellect in the world could not re-place the value of love and reason. Fromm was baffled by the many trends he observed as a scholar and researcher. He de-scribed himself as a Radical Humanist and described radical humanism as the philosophy that emphasizes the oneness of the human race, the capacity of each individual to develop his or her own powers and to arrive at inner harmony while help-ing to establish a peaceful world.

CARL JUNG (1875-1961)

Perhaps the most controversial of all of Freud's relationships is the one he maintained with Carl Jung. Like Rank and all the other followers, Carl Gustav Jung, got to know Freud through his writings since he was a prolific writer and wrote on a variety of subjects. He admired Freud and got a chance to meet him in Vienna in 1907. The story goes that after they met, Freud cancelled all his appointments for the day, and they talked for 13 hours straight, such was the impact of the meeting of these two great minds. Freud eventually came to see Jung as the crown prince of psychoanalysis and his heir apparent, but Jung had been entirely sold on Freud's theory and their relationship began to cool in 1909 during a trip to America. They were entertaining by analyzing each other's dreams, when Freud began to an excess of resistance to Jung's effort at analysis. Freud finally said they'd have to stop because he was afraid he would lose his authority. Jung felt rather insulted. The following account from Eric Fromm in his book Sigmund Freud Mission tells more about the relationship and gives further insight into Freud the man. Freud's dependency on the mother figure was not restricted to his wife and his mother. It was transferred to men, older ones like Breuer, contemporaries like Fliess and pupils like Jung. But Freud had a fierce pride in his independence and a violent aversion to being the protégée. This pride made him the awareness of dependency and negated it completely by breaking off the friendship when the friend failed in the com-

119

plete fulfillment of the motherly love. Thus, his great friendships follow the same rhythm, intense friendship for several years, then complete break, usually to the point of hatred. This was the fate of his friendship With Breuer, Fleiss, Jung, Adler, Rank and even Ferenczi, the loyal pupil who never dreamed of separating himself from Freud and his movement. Breuer, an older and successful colleague, had given Freud the seed of the idea, which was to develop into psychoanalysis. Breuer had been treating a patient, Ana O., and discovered that whenever he put her into hypnosis and made her tell him what was bothering her, she would feel relieved of her symptoms (depression and confusion).Breuer understood that the symptoms were caused by an emotional upheaval she had experienced while nursing her sick father, and furthermore he understood that the irrational symptoms were meaningful once one understood their origins. Thus, Breuer gave Freud the most important suggestion he ever received in his life, a suggestion which formed the basis of the central idea of psychoanalysis. Beyond that, Breuer acted toward Freud as a fatherly friend, including also considerable material help. How did this relationship end? True, there was a developing theoretical disagreement, because Breuer did not follow Freud in all this theory about sex. But certainly, such theoretical disagreement would not normally lead to a personal break, not to speak of the hatred Freud felt toward his former friend and benefactor. Or to put it in Jones' words: The scientific difference alone could not account for the bitterness with which Freud wrote about Breuer in his correspondence with Fleiss during the 1890s.

One valid criticism of Sigmund Freud is that he failed to describe anything positive about the unconscious. He strongly believed that the goal of psychotherapy was to reveal thoughts hidden in the unconscious or subconscious, but he made it sound so unpleasant that one had to think twice whether the efforts to go there were really worth it. He described it as a cauldron of seething desires, a bottomless pit of perverse and incestuous cravings, a virtual burial ground for frightening experiences which persistently came back to haunt us. Not exactly the kind of thing anyone looks forward to bringing into consciousness. In his earlier years Freud speaks of the oedipal complex and out of intellectual laziness many of the neo-Freudians find it difficult to transcend that narrow characterization of Him.

ROLLO MAY (1909 - 1994)

The circle of men responsible for what is loosely known as the Psychoanalysis Movement is completed with Rollo. He was the only American in that circle that lived to the ripe old age of 83, but his upbringing was less than pleasant, with parents divorcing when he was still a child and his sister suffered from severe psychosis. To make matters worse, May also had a serious bout with tuberculosis around the age of 30 that almost cost him his life. He spent three years in a sanatorium and while contemplating the possibilities of death, he threw himself into reading. The existentialist writings of Soren Kierkegaard became one of his favorite subjects. At the White Institute he met and became acquainted with Eric Fromm. Earlier on, upon graduation from College, he spent some time in Europe where he met, became acquainted with and was inspired by Alfred Adler. In 1953 May published his work entitled "Man's search for Himself". In it he offers some deep and accurate analysis into the psyche of modern man and his afflictions. The very preface of the book is vintage May: One of the blessings of living in an age of anxiety is that we are forced to become aware of ourselves. The painful insecurity on all sides gives us new incentive to ask, is there perhaps some important source of guidance and strength we have overlooked? People ask rather, how can one attain inner integration in such a disintegrated world? Or they question: How can anyone undertake the long development toward self-realization in a time when practically nothing is certain either in the present or in the future? Most thoughtful

122

people have pondered these questions. The psychotherapists have no magic answers. But there is something in addition to his technical training and his own self-understanding which gives an author the courage to rush in where angels fear to tread, thus offering his ideas and experience on these difficult questions. This something is the wisdom the psychotherapist gains in working with people who are striving to overcome their problems. He has the extraordinary, if often taxing, privilege of accompanying persons through their intimate and profound struggles to gain new integration. And dull indeed would be the therapists who did not get glimpses into what blinds people from themselves, and what block them in finding values and goals they can affirm. I do not see how the therapist can be anything but deeply grateful for what he is taught daily about the issues and dignity of life by those who are called his patients. Our aim is to discover ways in which we can stand against the insecurity of our times, to find center of strength within ourselves, and as far as we can to point the way towards achieving values and goals which can be depended upon in a day when very little is secure. One of the first things necessary for a creative relationship with others is to remove the subject of religious from the dialogue. He feels that making god an entity, a being over other beings, located in space is a carryover from a primitive view, full of contradictions and easily refutable. Religion or lack of it is shown not to some intellectual or verbal formulation but to one's total orientation to life. Religion is whatever the individual takes it to be. One's religious attitude is to be found at that point where he has a conviction that there are values in human existence worth living and dying for. The point we wish to emphasize is that psychologically religion is to be understood as a way of relating to one's existence.

These six giants of the field of psychology are presented to help familiarize students with the pioneers of the field. In keeping up with its commitment to bringing more students into the field of psychology, particularly at Saybrook University where I am putting the finishing touches on my own PhD, it seems natural to offer them a synopsis of the views of these great thinkers as they endeavor to take the field to higher heights and making it a more effective part of the life of those in need of assistance.

THE CONCEPT OF RACE

In his 1964 publication: *Man and his Ancestry,* Alan Houghton Brodrick commented:

> The fact is that the word race should be left to the *racist* and not used in speaking or writing about anthropological matters. There is only one race, and that is the human race. (p. 31).

This statement epitomizes everything we attempt to do at the Center for Intellectual Development. We add to this the fact that race is a social construct, an invention if you will, by those who traditionally play the role of divide and conquer. They succeed at this by confusing the issue and by confusing everyone within their purview. Confuse in order to weaken. We've declared therefore that the term race for defining different groups within the human family is dead. The more enlightened members of the species have taken a stance against the manipulation and division that this term represents

Africa as the birthplace of the human species is no longer a statement that is in dispute within the enlightened members of academia. The current dialogue is around migration, and the process through which geographic location affected skin tone, facial features and other physical characteristics.

In the United States for example, the people of African descent have now come to proudly identify themselves as African Americans. That is a major step from the early XX century when Negro, and Colored were the words of choice for self-description. They were often put down with more derogatory terms but that will not be a part of the conversation in this booklet. Having renounced color for defining themselves is what is needed for this group to take that strategy one step further by defining the other groups in similar technical terms. In doing so, they will come to the conclusion that there are no white people, no more than there are yellow people, or red people. Because of its historic implications, the term white has come to symbolize superiority. Remove the term and the psychological playing field is leveled even further. What they refer to as white people are simply European Americans who arrived in America from various parts of Europe with a traditional aggressiveness that continues to characterize their behavior. It is imperative also, for the purpose of empowerment that they take the time out to study the psyche and the behavior of those they refer to as European Americans. This will instantly reduce the influence this group has held over them since the Atlantic slave trade. It will also allow this terribly oppressed group to begin forging a new world, one that is free from the undue influence. There is no doubt that the internalization of the scientific reality described here will help that process.

OUT OF AFRICA

The great playwright Robert Ardrey wrote:

> *Not in innocence, and not in Asia was mankind born, the home of our fathers was that African highland reaching north from the Cape to the Lakes of the Nile. Here we came about – slowly, ever so slowly—on a sky- swept savannah glowing with menace. In neither bankruptcy nor bastardy did we face our long beginnings. Man's line is legitimate. Our ancestry is firmly rooted in the animal world, and to its subtle antique ways our hearts are still pledged. Children of all animal kind , we inherited many a social niceties as well as the predator's way. But most significant of all our gifts as things turned out, was the legacy bequeathed us by those killer apes, our immediate forebears. Even in the first long days of our beginnings we held in our hands the weapon, an instrument somewhat older than ourselves. Man is a fraction of the animal world. Our history is an afterthought, no more, tacked to an infinite calendar. We are not so unique as we should like to believe. And if man in a time of need seeks deeper knowledge concerning himself, then he must explore those animal horizons from which we have made our quick little march.*

(Ardrey 1963, p 9).

In the past few years scientist have given new meaning to the term out of Africa for even though it has been revealed for decades that Africa is the cradle of human kind and the birthplace of civilization, details on how the species migrated from there to the rest of the globe were not yet available.

The recent human genealogical project sponsored by National Geographic has already began to remove the mystery on how this development took place.

The study began and continues to be centered around an effort to trace mitochondrial DNA, the type of DNA that is traced only to the female of the species. This information is useful in giving the individuals a profile of their genealogical structures and helping them to predict some of the diseases they might be exposed to given their genealogical profiles. To get the job done they were compelled to produce a map of the migrations the species embarked on as they settled different territories around the globe. The deeply scientific complexities involved in DNA tracing is not something we are prepared to embark on in this manual but it is freely available to anyone who wish to educate themselves on this issue.

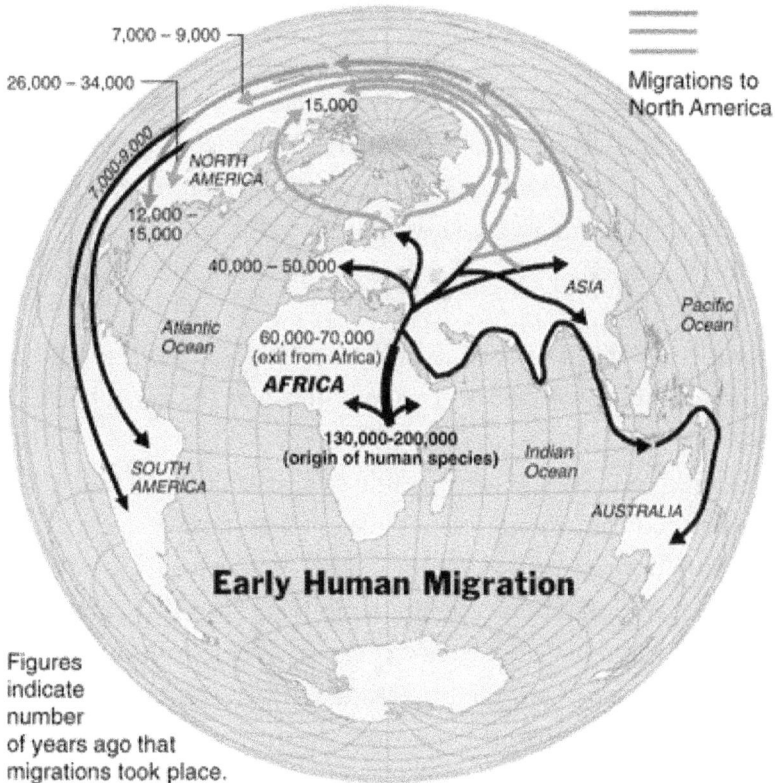

Early Human Migration

7,000 – 9,000

26,000 – 34,000

15,000

7,000-9,000

NORTH AMERICA

12,000 – 15,000

40,000 – 50,000

Atlantic Ocean

60,000-70,000 (exit from Africa)

AFRICA

130,000-200,000 (origin of human species)

SOUTH AMERICA

ASIA

Pacific Ocean

Indian Ocean

AUSTRALIA

Migrations to North America

Figures indicate number of years ago that migrations took place.

129

EDUCATION AS A LIFE PRESERVER

Also, at the time of this writing a psychological crisis is brewing among young African American males convinced that their lives can be snuffed out at whim by any European American law officer, but also by any European American with a gun. The only justification that needs to be presented for this is that they were in fear of their life.

This is a crisis that has mature African American males like myself scrambling for answers. The religious and the academic institutions are at a loss for answers as it relates to this crisis. As an organization, our only answer to this crisis for the time being is a renewed focus on education, not just conventional education with its road blocks and unnecessary delays, but a revolutionary approach to education that guarantees a motivated student a clear path to a PhD. There have not been any reported cases of a doctor of philosophy gunned down for any reason whatsoever regardless of the circumstances. For one, not only do they not put themselves in situations where they can be easily gunned down, but as professionals, they are better prepared to de-escalate any human situation, and fully prepared to handle crisis without emotions. No doubt this is a long-term plan, unlikely to save the life of the young man who is scheduled to be gunned down by a zealous cop tomorrow or a poisoned racist who will seize on the opportunity the current racist climate offers him or her to take a Black life with impunity.

Black boys are an endangered species at this moment in history, and measures must be taken to protect them.

A fast track towards a PhD takes care of several things at once: A- That child becomes a leader in training, one whose open-mindedness, emotional stability and intellectual fortitude will positively influence those around him.

B- It immediately reduces the rate of poverty by one.

C- It is one less person for the prison system in the United States.

This fast track to a PhD does not exclude girls or young people of other ethnic groups, they must however, understand the premise of the program, and must be willing to contribute intellectually to the remedying of this crisis.

Given the mission of this organization the curriculum for these degrees leans heavily towards science and psychology, and that is because these two areas of study allow its professionals to bring clarity and direction to those in its search. By their very nature, and for the mission that is embarked on, these two areas of study that are wide in their scope, offer job security for nearly everyone involved. The hate, confusion, misinformation, depression and low self-esteem among the masses in our times almost guarantees this.

AN ENCOURAGING OUTLOOK

If the species is to advance, intellect will have to play a greater role than it does today. Much has been said about the fancy gadgets children are exposed to today and their extraordinary abilities to master these gadgets in zero time, but some doubts remain as to whether this has had much effect on the intellectual growth of these children, and their sense of independence as individuals. Although there are hardly any signs of organized resistance on the horizon, it is hard to imagine the species holding steadfast to these discredited belief systems for much longer, but highlighting the damage they continue to cause on young minds appears to be perhaps the one inevitable way of confronting a structure that has been so effective in slowing human growth and bringing so much misery to so many over the ages.

As we embark on this quest to understand human behavior, particularly as it is affected by belief or belief systems, it is important to recognize that for reasons beyond our explanation, there is precious little data available on this subject. One gets the impression that the power to ostracize those who dare to openly criticize the phenomenon of belief remains an ever-present deterrent for scholars who by their own nature have resisted the power and control held by religions of all sway even in our era of modernity. The data available on belief and its effect on human behavior, particularly its effects on human brutality is scant, a factor that causes this research to be all the more challenging, for despite its importance and the enormous effect belief has had on the human psyche, as well as on human behavior, scholars everywhere have avoided this issue at all cost. This seemingly collective decision; at least at a subconscious level; has had the effect of leaving the masses in every culture and in every society, up to their own devices. Since the brightest among them avoid the subject, they are left to figure out this subject on their own despite the confusing quandary in which nearly all happens to find themselves in, particularly as it relates to this subject.

It is interesting that Harris *et al* (2009) would argue that the industrial world anticipated the demise of religion as we know it given the extraordinary promises of technology and a more advance way of thinking as it was expected. For reasons that are nearly impossible to explain, this turned out not to be the case. This prompt the researchers to delve into the phenomenon of belief and the mechanisms involved in the stranglehold it has succeeded in maintaining over nearly all of those who refer to themselves as humans. According to Harris *et al* (2009) there appears to be a continuous correlation between culture and religion although the authors fail to give us any definition for either of those terms. It behooves us then to pause and attempt to identify these terms before we advance in this discussion thus I shall refer to religion as the process through which human beings attempt to identify with the forces responsible for their existence, and for that same purpose I shall define culture as "learned behavior." As we proceed with these simple definitions we recognize that the guess work has been taken out of the subject.

Norman et al (2008) tells us that the human mind possesses a limited capacity for processing information, and it may be fair to say that this may just be the central point of why dogma and belief in the hands of skillful manipulators, have been able to maintain such grasp on humans everywhere. The question to be asked is, at what point do we allow irrationality to govern the affairs of humans? In describing the general penchant for fiction that appears to run common in all humans Norman et al (2008) explains that we embrace only a portion of the information that is fed to us by way of fiction, that even though at first, we appear to believe it all, as some degree of rationality imposes itself we begin to drop some of what is just too obviously ludicrous.

ADDICTION TREATMENT
AND THE POLITICS OF DRUGS

A Non-Governmental Organization committed to empowering its members intellectually would not be complete without an analysis of the impact of illicit drugs, not just in our society but in the world we inhabit. Illicit drugs impact those we serve so it is inevitable that we take the time out to do a careful analysis of this subject, and define the attitude our organization intends to maintain on the subject. It goes without saying that the widespread violence produced by the illicit drug trade is not limited to the United States, here children caught in the cross fires of gang wars are being killed nearly every day, but the scenario appears to be even worse in countries south of the United States and the Caribbean where young people compete for the opportunity to get their products to the drug abusing market that exists in the United States. A necessary component of intellectual pursuit is recognizing the reality of the social environment, for despite our efforts to promote intellectual growth and intellectual empowerment, recognizing the allure of the lucrative business of drug trafficking represents, in many cases a nearly insurmountable hurdle in our efforts to reach a good many adolescent. This reality compels us to revisit and discuss ideas like de-penalization, de-criminalization and even legalization as it relates to drugs, ideas that can only be discussed adequately by individuals with advanced intellect.

An additional damage caused by our policy towards drugs is that human beings seeking to escape the crime ridden areas of countries south of the United States, will head for the only country they can gain access to for their mere economic survival, but when they pursue that only option they are immediately made criminals by virtue of their survival effort. When they're caught and eventually sent back to their country of origin, made uninhabitable by the trafficking and violence they originally attempted to escape, desolation, desperation and crime often becomes their only survival option.

As we contemplate the three words mentioned earlier: de-penalization, de-criminalization and legalization, it may help to take a look at the societies that have experimented with these concepts. We've all become familiar with the case of the Netherlands where are reported to be using at will, the kinds of drugs that are illegal in the United States. The opinions vary widely as to the actual results of the Netherlands experiment in drug accessibility, but a more recent experiment with drug accessibility was put into practice in Portugal, highly controversial in nature and with a great deal of predictions regarding its impending failure. Yet in an article published in the April 7, 2009 issue of *Scientific America* entitled, Five Years After: Portugal's Drug Decriminalization Policy Shows Positive Results, writer Brian Vastag argues that since the beginning of the experiment in 2001street drug related death from drug overdose fell significantly, as did the cases of new HIV infection for in the face of a growing number of deaths and cases of HIV linked to drug abuse, the Portuguese government in 2001 tried a new tack to get a handle on the problem—it decriminalized the use and possession of heroin, cocaine, marijuana, LSD and other illicit street drugs. Five years later, the number of deaths from street drug overdose dropped from around 400 to 290 annually, and the number of new HIV cases caused by using dirty needles to inject heroin, cocaine and other illegal substances plummeted from nearly 1,400 in 2000 to about 400 in 2006. The focus was on treatment and prevention instead of jailing users would decrease the number of deaths and infections, so instead of being put into prison, addicts are going to treatment centers and they're learning how to control their drug usage or getting off drugs entirely. Vastag (2009) then goes on to give some details of the Portuguese experiment, telling us that under the 2001 law, penalties for people caught dealing and trafficking drugs are unchanged; that dealers are still jailed and subjected to fines depending on the crime. But people caught using or possessing small amounts—defined as the amount needed for 10 days of personal use—are brought before what's known as a "Dissuasion Commission," a special administrative body made up of three persons. That there were several of these commissions each including at least one lawyer or judge and one health care or social worker. As it stands, Vastag (2009) tells us, the panel has the option of recommending treatment, a small fine, or no sanctions whatsoever. Vastag (2009) then cites Walter Kemp, a spokesperson for the United Nations Office on Drugs and Crime, stating that decriminalization in Portugal "appears to be working and that his office is putting more emphasis on improving health outcomes, such as reducing needle-borne infections. Drug legalization, argues Vastag (2009), removes all criminal penalties for producing, selling and using drugs; no country has tried it. In contrast, decriminalization, as practiced in Portugal, eliminates jail time for drug users but maintains criminal penalties for dealers. Spain and Italy have also decriminalized personal use of drugs and Mexico's president has proposed doing the same (p. 1). The August 29, 2009 issue of *Economist*, published an article entitled: Treating, not punishing. A specific author was not mentioned for the article but it made continuous reference to a study done by *constitutional lawyer* Glenn Greenwald regarding the Portuguese experiment with drug decriminalization. In it Greenwald is said to be arguing that the evidence from **Portugal** since 2001 is that decriminalization of **drug** use and possession has many benefits and no harmful side-effects, despite the fact that in 2001 newspapers around the world carried graphic reports of addicts injecting heroin in the grimy streets of a Lisbon slum. When a young British backpacker was found comatose on a Lisbon street corner, the government took action, and the result was a sweeping decriminalization law. The personal use and possession of all drugs, including heroin and cocaine was allowed. The foreign media expressed concerns that holiday resorts would become dumping-grounds for **drug** tourists. Some conservative politicians went as far as denouncing it as lunacy, convinced that plane loads of foreign students would head for the Algarve to smoke marijuana. The report then assesses Portuguese drug policy in the context of the EU's approach to drugs. The varying legal frameworks, as well as the overall trend toward liberalization, are examined to enable a meaningful comparative assessment between Portuguese data and data from other EU states. Greenwald's conclusion was that judged by virtually every metric, the Portuguese decriminalization framework has been a resounding success, and that within this success lie self-evident lessons that should guide drug policy debates around the world (p. 1).

Even the United Nation has weighed in on the Portuguese experiment. In an article appearing in the June 24, 2009 edition of the *Huntington Post,* writer Ryan Grim commented that, in an about face, the United Nations praised drug decriminalization in its annual report on the state of global drug policy. This is significant because in previous years the UN drug *czar* had expressed skepticism about Portugal's decriminalization, which had removed criminal penalties in 2001 for personal drug possession and emphasized treatment over incarceration, and that the policy was in violation of international drug treaties, but after a mission to Portugal in 2004 the International Narcotics Control Board noted that the acquisition, possession and abuse of drugs had remained prohibited, and commented that the practice of exempting small quantities of drugs from criminal prosecution is consistent with the international drug control treaties. Grim (2009) commented that the Executive Director of the United Nations Office of Drugs and Crime Antonio Maria Costa has been exploring the debate over repealing drug controls, acknowledging that attempts to control the flow and use of drugs have generated an illicit black market of macro-economic proportions that uses violence and corruption as its primary tools. Grim (2009) added that despite Costa's openness on the subject, Jack Cole, executive director of Law Enforcement Against Prohibition (LEAP) and a retired undercover narcotics detective, objected to the report's classification of current policy as *control* commenting that the world's 'drug czar,' Costa would have you believe that the legalization movement is calling for the abolition of drug control, but the opposite is true he added, we are demanding that governments replace the failed policy of prohibition with a system that

139

actually regulates and controls drugs, including their purity and prices,   as well as who produces them and who they can be sold to. You can't have effective control under prohibition, as we should have learned from our failed experiment with alcohol in the U.S. between 1920 and 1933.

CLUB VIZCAYA INTERNATIONAL

It is at this point in the writing that we must warn you that the Center for Intellectual Development is not just a gloom and doom organization taking on the evils of society; on the contrary, it is a fun filled organization committed to the health and happiness of its members. Teaming up with Club Vizcaya International an organization managed by my wife Yolanda, clients receive a taste of both worlds. In one they grow intellectually, become fortified emotionally, and learn the reality about the world. In the other they are set a loose to carry on and have fun in healthy environment. She organizes the parties in which members of both the club and the organization are pampered and catered to as they make claim of their right to happiness. Healthy eating and moderate drinking are encouraged at these fun-filled activities, and members come away with the notion that the world is theirs for the taking. A unique building program has members visualizing the possibilities of owning their homes outright, not in partnership with any bank or manipulating financial institution. An intense *Cardiosalsa* and a unique approach to cooking and dining keeps members healthy and happy as they in turn work to improve the lives of others. This is what we refer to as a win situation.

Vizcaya communities is the upscale housing program championed by Club Vizcaya International, and it rounds out the actual intentions of this movement which is to improve people's life by reducing for them the cost of living. That of course, is only for members, and membership is available to everyone providing they're able to follow some basic behavior rules.

141

The open-air concerts and other events planned by Club Vizcaya International are labeled: ***Celebrating Humanity***. Most if not all of them will be fundraisers for targeted causes, and chief among those causes is child illness. Members of our organization are committed to frequent visits to children's hospital to bring joy where they can and to remain grounded as it relates to their own good fortune.

CARDIOSALSA

If we pretend to improve the lives of those who choose to be part of this process it is necessary to start with health, it does not serve us if the health of all those operating in this organization have their health compromised. Both the Center for Intellectual Development and Club Vizcaya International are committed to the health of all those who operate within their ranks, and for this we have developed an extensive health program that begins first with adequate information on everything the individual is required to do in order to preserve his or her health.

As an effective support network, we recognize how difficult it is for the individual to remain motivated to take and maintain all the actions needed to remain healthy. That is why we take persuasiveness and emotional support seriously, seeing it as crucial elements in achieving the goals the individual sets out for him or herself. Health experts working in this organization will ensure that every member receives the assistance needed to Achieve their goals.

To maintain the health of all, physicians recommend that the individual keeps moving and that, we recognize, requires a certain degree of motivation. One of the most effective ways to keep moving is through dance and where dance is concerned there are few that meet that purpose better than the dance form we've come to know as *salsa*. It is a way of dancing that gives instant happiness to its practitioners, and that allows them to practice it more frequently and for more extended periods, providing the environment is healthy. The events at Club Vizcaya International fulfill these goals, as they provide members with a safe environment, and offer them the opportunity to exercise for extended hours in accordance with the recommendations of their physician, health fitness consultants and their nutritionist.

Our own nutritional programs provide members with the most advanced information on the methods that exist for the maintenance of health. Members receive information on free radicals, on antioxidants, on hydrogenation of fats, on the abuse of refined carbohydrates, infections, inflammations and on cell deterioration.

The cost of participation in an advanced program like ours is affordable to everyone so we invite you to participate.

CONCLUSION

There is not much that can be said by way of conclusion on a subject as complex as the one dealt with here, so at the risk of boring the reader I will attempt to end on a high and hopefully more encouraging note.

Strange as it may sound, what is proposed on these pages is a new approach to psychology, the psychotherapeutic process, and to human relation in general. How far this proposed new approach reaches in anyone's guess but I have committed the remainder of my time on this planet to see to it that this proposal takes root, and that those who wish to benefit from this new approach can do so without reservations or limitations. Fortunately, it does not require a new structure, it simply builds on what is already in existence, adding a twist to it. That twist is brutal honesty regarding not just the birth of our planet, but also the beginning of life on it and everything that was to come thereafter. As simplistic as this may appear, those capable of handling this approach are few, and they are far in between. That said, the process is already on its way, and all of it is being shouldered by the Center for Intellectual Development, the organization I happen to be heading. The name was chosen in order that there be no doubt whatsoever regarding what the organization stands for.

The vigorous application of the intellect is a practice that is greatly discouraged in our world, and there are those who believe that this is by design. I have no opinion on whether this is meant to be, I know however, that for one to participate within these ranks they must be willing to apply the intellect.

This is also not an attempt to separate thinkers from non-thinkers but rather a determination to exclude those who have given up their rights to think and to function at a higher level intellectually.

The organization is unconcerned with numbers or statistics, all that it concerns itself with is that those who choose to function at a higher level find a place that encourages such determination, and that they succeed in their goal to make this a better species.

Understanding the trajectory of the human species is a key component of emotional stability and mental health. That is the premise behind the psychotherapeutic method introduced in these pages.

Bibliography

Ardrey, Robert. (1963). *African Genesis: A Personal Investigation into the Animal Origins and Nature of Man.* New York N.Y. Delta Books.

Arianrhod, R. (2012). Seduced by logic: Emilie du Châtelet, Mary Somerville, and the Newtonian revolution (US ed.). New York: Oxford University Press. ISBN 978-0-19-993161-3.

Aries, Phillip. (1960). *Centuries of Childhood: A Social History of Family Life.* New York, N.Y. Vintage Books.

Barrow, R. Bailey, R. Carr, D., & McCarthy, C. (2010). *The SAGE handbook of philosophy of education.* Thousand Oaks CA. Sage Publication.

Benoit Denizet-Lewis Times Mag June 25, 2006 Horacio Salinas. Online at: www.nyti-mes.com/2006/06/25/magazine/25addiction.html

Blau, T. H. (1988). *Psychotherapy tradecraft: the tech nique and style of doing therapy,* Bristol Pa. The Taylor Francis Group.

Bodanis, D. (2009). Passionate Minds: The Great Love Affair of the Enlightenment. New York N.Y. Crown. ISBN 0-307-23720-6.

Borden, W. (1999). *Comparative approaches in brief dy namic psychotherapy.* Binghamton N.Y. Haworth Press.

Bradley, K., & Cartledge, P. (2011). *The Cambridge world history of slavery: Volume 1.* New York NY

Cambridge University Press.

Breuer, J. Freud, S, and Luckhurst, N. (2004). *Studies in Hysteria*. New York. Penguin Books.

Bringuier, J.C., & Piaget, J. (1977). Conversations with Jean Piaget. France Editions Laffont, S.A

Brodie, F. (1971). *No man knows my history: The life of Joseph Smith, the Mormon prophet.* New York N.Y. Alfred A. Knopf Publishers.

Bushman, R., & Woodworth, J. (2005). *Joseph Smith: Rough stone rolling.* New York N.Y. Alfred A. Knopf Publishers.

Carlisle, R. (1975). *The Roots of black nationalism.* Port Washington NY; National University Press.

Carter, R. (1999). *Mapping the Mind.* Berkeley and Los Angeles. University of California Press

Colaiaco, J. (2006). *Frederick Douglass and the Fourth of July.* New York NY: Palgrave Macmillan.

Cooper, B. (2013). Alan Turing: His Work and Impact. New York N.Y: Elsevier. ISBN 978-0-12-386980-7.

Copeland (ed.), B. Jack (2005). Alan Turing's Automatic Computing Engine. Oxford: Oxford University Press. ISBN 0-19-856593-3. OCLC 224640979.

deMause, Lloyd. (1974). *The History of Childhood.* New York, N.Y. The Psychohistory Press.

deMause, Lloyd. (1992). *The History of Child Abuse.* The Journal of Psychohistory, 25. (3).

de Mause, Lloyd. (2002). *The Emotional Life of Nations.* New York. Other Press LLC.

Despert, L. (1970). *The Emotionally Disturbed Child:*

An Inquiry Into Family Patterns. Garden City N.Y.
Double Day & Company, Inc.

Dewey, J.(1900). *The School and Society.* Chicago Ill.
The University of Chicago Press.

Dewey, J. (1902). *The School and Society.* Chicago IL The
University of Chicago Press.

Dewey, J. (1922). *Democracy and Education: An intro*
Point Books

Dodds, A. (2009). *The Abrahamic faiths? Continuity
and discontinuity in Christian and Islamic doctrine.
Evangelical Quarterly,* 81 (3), pp 24, 230, 253.

Douglas, F. (1852). *Fourth of July Speech.* [Online]
www.freemaninstitute.org

Du Bois, W. E. B. (1903). *The souls of black folks.* Chicago IL: A.C. McClurg & Co.

Duncan, B. Miller, S, and Sparks, J. (2004). *The Heroic
Client, A Revolutionary Way to Improve Effectiveness Through Client-Directed, Outcome-Informed
Therapy.*San Francisco, CA. Jossey-Bass, Publishers.

Ellis, A. (No date). *The Essence of Rational Emotive
Behavioral Therapy.* Available (Online):
http://www.rebt.ws/albertellisbiography.html

Ellis, A. & Dryden, W. (1997).*The Practice of Rational
Emotive Behavior Therapy.* New York Springer
Publishing Company.

Ellis, A. & Blau, S. (1998). *The Albert Ellis reader: a
guide to well-being using rational emotive behavior
therapy.* New York. Kensington Publishing Corpo-

ration.

Ellis, A. (2002). *Overcoming Resistance*. New York N.Y.
    Springer Publishing Company.

Erikson, E. (1980). Identity and the life cycle, Volume 1.
    New York   W.W. Norton & Company, Inc.

Frank, J.D. & Frank, B. (1961,1973, 1991). *Persuasion &
    Healing.* The John Hopkins University Press

Frankl, V. (1992). *Man's search for meaning: an introduc
    tion to logo-therapy.* Boston MA. Beacon Press.

Gates, H. L. (1994). *Frederick Douglass autobiographies.*
    New York NY: Literary Classics of The United
    States.

Genovese, E.(1972). *Roll jordan roll, The world the slaves
    made.* New York, NY: Random House Inc.

Green, D. (1998). *Hidden lives: voices of children in Latin
    America and the Caribbean.* London. Cassell
    Wellington House.

Harlan, L. (1983). *Booker T. Washington: The wizard of
    Tuskegee.* New York. NY: Oxford University Press

Harris, S. (2005). *The end of faith: religion, terror, and the
    future of reason.* New York N.Y. W. W. Norton &
    Company.

Harrison, V. (2006). Scientific and religious worldviews:
    antagonism, non-antagonistic incommensurability
    and complementarity. *Heythrop Journal, 47 (3). P.
    18, 349-366.*

Haugen, Brenda (2006). *Joseph Stalin: Dictator of the
    Soviet Union.* Minneapolis MN. Compass Point

Books.

Ibáñez, F. (1964). Tales of Philosophy. New York N Y. Clark-
son N Potter, Inc/Publisher.

Ikeda, D. (2001). *Soka Education: A Buddhist Vision for
teachers, Students and Parents* Santa Monica, CA.

Jessop, F. &, Brown, P. (2010). *Church of lies.*
San Francisco CA. Jossey-Bass.

Journal of the American Medical Association. December
25, 2002 288,. 24, pp. 3096-3101Middleway Press.

Kramer, R. (1976). *Maria Montessori: a biography.*
Chicago Ill. University of Chicago Press.

Levering, D. (2000). *W.E.B Du Bois—the fight for equality
and the American century, 1919-1963.* New York,
NY: Henry Holt and Company, LLC.

Lloyd, P. & Fernyhough, C.(1999). *Lev Vygotsky:
critical assessments, Volume 1.* New York
Rutledge.

Loentz, Elizabeth. (2007). *Let me continue to speak the
truth: Bertha Pappenheim as author and activist.*
Jerusalem Hebrew Union College Press.

Manning, S. (2004). *Psychology, symbolism, and the sacred: Confronting religious dysfunction in a changing world.* Otsego MI: Page Free Publishing, Inc

Mc Ginn, L. (1997). American Journal of Psychotherapy. 51
(3), pp 309, 8

Miller, A.(1998). *The Political Consequences of Child
Abuse.* The Journal of Psychohistory, 26. (2).

Miller, A., & Jenkins, A. (2010). *Free from lies: Discover ing your true needs.* New York NY: W.W. Norton and Company Inc.

Miller, W. R. & Rollnick, S. (1987). *Motivational inter viewing: preparing people for change.*

Miller, W. R. (1994). *Motivational enhancement therapy manual: a clinical research guide for therapists treating individuals with alcohol abuse and dependency Volume 20*

Montessori, M. & Wyman, H.(1912). *The Montessori Method: Scientific pedagogy as applied to children in education.* New York Frederick A. Stokes company

Montessori, M.(2004). The Discovery of the Child. Delhi India. Aakar Books.

Mooney, C.G.(2000). *Theories of Childhood. An Introduction to Dewey, Montessori*

Moore, J. (1965). *Booker T. Washington, W.E.B. Du Bois, and the struggle for racial uplift.* Wilmington, DE. Scholarly Resources Inc.

Payne, G. H., & Jacobi, A. (1916). *The Child in Human Progress.* New York. G. P. Putnam's Sons.

Pussaint, A. (2000). *Lay My Burden Down: Suicide and the Mental Health Crisis Among African Americans.* Boston MA. Beacon Press.

Woodson, C.G. (1933). *The mis-education of the Negro.* Washington, DC: The Associate Publishers Inc.

Reuter, C. (2004). My life is a weapon: The history of suicide bombing. Princeton N.J Princeton University Press.

Riak, J. (2009). *Plain Talk About Spanking*. Available
 [Online] http://www.nospank.net/pt2009.htm
Rieber, R. and Robinson, D. (2001). *Wilhelm Wundt*
 In history: the making of a scientific psychology.
 New York Plenum Publishers.
Robinson, B.A. Jul 25, 2004 *Ontario consultants on Reli-*
 gious Tolerance Retrieved 1/11/11
 http://www.religioustolerance.org/flds.htm
Roekeach, M. (1960). *Understanding human values: Indi-*
 vidual and societal. New York, NY: The Free Press.
Scott, E. & Stowe, L. (1910). *Booker T. Washington,*
 builder of a civilization. Cambridge, MA: Andover-
 Harvard Theological Library.
Spindel, Donna J. *Assessing Memory,* Twentieth Century
 Slave Narratives Reconsidered Journal of Interdis-
 ciplinary History, Vol. 27. No. 2 (Autumn, 1996).
 Pp.247-261. Published by: MIT Press
Straus, Murray A. Donnelly, Denise. *Beating the devil out*
 of them: corporal punishment in American families
 and its effects on children. New Brunswick.
 Transaction Publishers.
Strauss, Murray. (2009). University of New Hampshire
 Children Who Are Spanked Have Lower IQs, New
 Research Finds. *Science Daily*. Available (Online)
 http://www.sciencedaily.com
United Nations *Declaration of the Rights of the Child.*
 Availble[Online]www2.ohchr.org/English/law/crc.htm.
Verhellen, E. (1996).*Monitoring Children's Rights*.
 The Hague, the Netherlands. Kluwer Law International.

Wall, E. &, Pulitzer, L.(2009). *Stolen innocence: My sto-ry of growing up in a polygamous sect, becoming a teenage bride, and breaking free of Warren Jeffs.* New York N.Y. Harper Collins.

Waite, R. (1971). Adolf Hitler's Guilt Feelings: A Problem in History and Psychology *The Journal of Interdisciplinary History, Vol. 1, No. 2, pp. 229-249*

Wampold, B. E. (2001). *The great psychotherapy debate: models, methods, and findings.* Mahwah N.J. Law rence Earlbaum Associates Publishers.

Williams, E. (1944). *Capitalism & Slavery* Chapel Hill, NC: The University of North Carolina

Williams, C. (1987). *The destruction of black civilization, Great issues of a race 4.500 B.C to 2000 A.D.* Chi cago IL: Third World Press.

Woodson, C. (1933). *The miseducation of the Negro.* San Diego, CA: The Book Tree Publishers

Wundt, W. M. (1904). *Principles of physiological psy chology, Volume 1.* New York. The Macmillan Co.

Zelizer, Viviana. (1985). *Pricing the Priceless Child: The Changing Social Values of Children* Princeton N.J. Princeton University Press.

Frye, D. (2004). *Save the women and children: An analy sis of the rhetoric in the social movement against FGM* (Order No. 1420791). Available from ProQuest Disser-tations & Theses Global. (305036207). Retrieved from https://tcsedsys-tem.idm.oclc.org/login?url=https://search-proquest-

com.tcsedsystem.idm.oclc.org/docview/305036207?ac-
countid=34120

www.ingramcontent.com/pod-product-compliance
Lightning Source LLC
Chambersburg PA
CBHW020002290326
41935CB00007B/279